WINES OF THE LOIRE VALLEY

THE CLASSIC WINE LIBRARY

There is something uniquely satisfying about a good wine book, preferably read with a glass of the said wine in hand. The Classic Wine Library is a series of wine books written by authors who are both knowledgeable and passionate about their subject. Each title in The Classic Wine Library covers a wine region, country or type and together the books are designed to form a comprehensive guide to the world of wine as well as an enjoyable read, appealing to wine professionals, wine lovers, tourists, armchair travellers and wine trade students alike.

Other titles in the series include:
Wines of the Rhône, Matt Walls
Côte d'Or: The wines and winemakers of the heart of Burgundy, Raymond Blake
The wines of Chablis and the Grand Auxerrois, Rosemary George
Wines of the Languedoc, Rosemary George MW
Madeira: The islands and their wines, Richard Mayson
The wines of Portugal, Richard Mayson
Sherry, Julian Jeffs
The wines of Austria, Stephen Brook
The wines of Australia, Mark Davidson
The wines of Piemonte, David Way
Rosé: Understanding the pink wine revolution, Elizabeth Gabay MW
Amarone and the fine wines of Verona, Michael Garner
The wines of Greece, Konstantinos Lazarakis MW
The wines of northern Spain, Sarah Jane Evans MW
The wines of New Zealand, Rebecca Gibb MW
The wines of Bulgaria, Romania and Moldova, Caroline Gilby MW
The wines of Germany, Anne Krebiehl MW
The wines of Georgia, Lisa Granik MW
The wines of Southwest USA, Jessica Dupuy
The wines of South Africa, Jim Clarke

WINES OF THE LOIRE VALLEY

BEVERLEY BLANNING MW

ACADEMIE DU VIN LIBRARY

Beverley Blanning is a London-based writer with extensive experience of the wines of the world. Author of *Wine Tasting* (teach yourself) and *Biodynamics in Wine* (IWFS), she has also contributed to the *Oxford Companion to Wine* and written for numerous publications. She has judged at a host of wine competitions around the world, (including in France, Portugal, Chile, Argentina, Turkey and the UK), and has been a member of the Decanter World Wine Awards panel since its inception more than 20 years ago. Recruited by the late Steven Spurrier to lecture on the legendary Christie's Wine Course in London, she now hosts private tastings and corporate events. She passed the Master of Wine examination at her first attempt in 2001 and has been an active member of the Institute of Masters of Wine ever since. She spends much of her spare time in the Loire Valley and can be found on Instagram at @bevblanningmw.

In loving memory of Jane MacRae
and a birthday under the stars in Anjou.

First published in 2024 by Académie du Vin Library Ltd
academieduvinlibrary.com

A CIP catalogue record for this book is available from the British Library
ISBN 978–1–913141–90–5

Publisher: Hermione Ireland
Editor: Rebecca Clare
Indexer: Kim Stringer
Illustrator: Darren Lingard

Printed and bound in Great Britain

CONTENTS

List of maps

PREFACE

As I sit down to write this, a mail drops into my inbox, offering a mixed case of wines from 'France's most exciting wine region'. No prizes for guessing which one it is. The Loire Valley is enjoying a renaissance in interest that some would argue is long overdue. And it goes well beyond wine merchants trying to sell cases.

Writing in the UK's *Guardian* newspaper earlier this year, David Williams described the Loire as not only the most exciting, but also the most dynamic region in France this century. More recently, in the *Financial Times*, drinks writer Alice Lascelles described Anjou, the largest region of the Loire, as 'a fulcrum for winemaking at its most radical'. Closer to home, the French newspaper *Le Figaro* has been unusually fulsome in its praise of Loire wines, boldly announcing in 2024: 'Muscadet is the new Montrachet'.

I've been visiting the Loire Valley for over twenty-five years and, from what I see today, the wines have never been better. In the early 2000s, unbridled enthusiasm for the Loire was thin on the ground. If any journalist had tried to compare Muscadet to Montrachet in a national newspaper, they would probably have lost their job (and possibly their wine loving friends, too). Today, though, there is a genuine excitement and energy permeating the cellars of the Loire, as more and more people are waking up to this region of largely unknown treasures just waiting to be discovered – and yes, that does include Muscadet.

So why has it taken so long for the Loire to rise above its stubbornly under-the-radar status? Winemaking here is hardly a recent pursuit and has long been recognized as an integral part of the rich historical landscape of the region. The answer is as complicated as the history of France itself, a story that involves fortunes gained and lost and appellations that have gone from heroes to zeros in less than a human lifetime. It is barely a century since the French vineyard was entirely replanted following the devastation of phylloxera, and the Loire – a vast, spread-out and varied winegrowing region without a clear, single identity – perhaps suffered more than most in finding its way in the aftermath of the crisis. Some places fared better than others: it was a stroke of genius on the part of the first Sancerre vigneron who had the idea of replacing difficult-to-manage Pinot Noir vines with the more productive Sauvignon Blanc. It turned out to be a terroir match made in heaven that is still the most successful wine style of the Loire today.

Elsewhere, however, growers suffered during the last century from attempting to follow fashion, often planting grapes to produce styles that were perceived to be in demand, rather than those best suited to their terroirs. As in many places in France, in the years following the Second World War the emphasis was on selling quantity, rather than focusing on quality. The advent of mechanization and chemical fertilizers did nothing to improve this situation in the years that followed.

But this story has a happy ending. The Loire now has one of the highest rates of organic certification in France (including the majority of the more than 100 domaines profiled in this book). For the quality-focused vignerons today the goal is not volume production, but the faithful expression of their terroirs. Ongoing studies are revealing more of the complexities of the region's soils, resulting in a proliferation of site-specific cuvées the length and breadth of the region. The Loire has also proved to be a magnet for new vignerons looking to create innovative wines with a natural focus, bringing youthful energy and alternative perspectives to what has long been considered a traditional winegrowing region.

As well as changing views on viticulture, there has also been a noticeable shift in winemaking practices, with greater respect for the quality of the fruit, and often a more hands-off approach in the cellar. These adaptations have led to a new wave of bright, modern, terroir-driven wines that major on subtlety and finesse. These are qualities increasingly hard to find in a warming world, but at which the Loire excels. For a diverse choice of wines with refreshment at their core, the Loire is hard to beat. And nowhere on Earth can match the Loire for the quality of wines from its signature varieties: Chenin Blanc, Cabernet Franc, Sauvignon and Melon. The best of these are truly outstanding wines that every wine lover should know.

The aim of this book is two-fold. First, it is intended as a reference for understanding the geography, appellations and wines of this large and complex region. The Loire can be difficult to get to grips with simply because there is so much of it: there are so many appellations, and so many grape varieties and styles of wine. There are also many thousands of growers, so the second aim of the book is to highlight some of these individuals and show the range and depth of talent in the Loire Valley today. This is by no means an exhaustive list (if it were, the book would never have reached completion), but I hope the personal selection here will encourage you to seek out these wines and motivate you to make discoveries of your own.

I also hope this book might encourage readers to travel to experience the wines and beauty of the Loire Valley for themselves. Included are maps of the key winegrowing areas, contact details for growers and clearly marked sections on suggested places to stay and eat. I can recommend the excellent food, wine and hospitality of the aptly named Garden of France. It has been a great privilege to spend time with the unfailingly kind winegrowers of the Loire Valley in researching this book. They are the heroes and heroines of the pages that follow, and I hope you enjoy reading their stories.

A note on the text

Unless otherwise stated, figures for hectares in production and volumes of wine in each appellation are the five-year averages, from 2019–2023, sourced from InterLoire. This presentation is chosen to give a more accurate representation of typical production, since vintage variation dictates that volumes can very significantly from year to year, affecting some appellations more than others. The most recent (2023) figures for production in each appellation are included in Appendix III, on p. 334.

Producers are listed alphabetically, and under the appellation where the winery is located.

Producers certified organic are marked ; those also certified biodynamic are also marked .

1

TWO THOUSAND YEARS OF WINE GROWING

Wine has long been an integral part of the cultural history of the Loire Valley. In 2000, UNESCO declared a 280-kilometre stretch of the river a world heritage site, describing it as 'an outstanding cultural landscape' that 'bears witness to … a harmonious development of interactions between humankind and their environment over two millennia.'

The first written evidence of vineyard cultivation was in the sixth century, by Grégoire, Bishop of Tours, when he made reference to the areas we know today as Vouvray and Chinon. Several centuries earlier, in 77 CE, Pliny the Elder had noted in his *Natural History* that there were vines growing on the banks of the Loire, but it was only in 2016 that archaeologists discovered wine presses and signs of amphora production (which would necessarily have been in proximity to the vineyards), confirming the presence of winemaking in Anjou, Bourgueil and other parts of Touraine as far back as the first to third centuries CE.

We don't know the nature of these early wines, but it is possible that Cabernet Franc was among the varieties cultivated. At around the turn of the first century CE the Gallic tribe known as the Bituriges, based around the town of Bourges, occupied a large territory covering much of the central portion of today's Loire vineyards, from Sancerre to Haut-Poitou. The tribe had commercial links with Bordeaux from where, it is suggested, they sourced plants for the purposes of wine production. These were described in Pliny's writings as Biturica, an unproven but possible ancestor of Cabernet Franc.

In the Middle Ages, it was the monks who fostered the development of the vine in many parts of the Loire, including Saint-Pourçain, Sancerre and Angers. Wine was considered essential, not just for its religious significance in the daily life of the abbeys, but also for receiving the sick, pilgrims and passing travellers, and for donors to the church. This also favoured production of wine close to the towns and villages surrounding these religious centres.

The fifteenth and sixteenth centuries could be considered a high point for the reputation of Loire wines. Not only was the Loire the hub for French royalty, which greatly enhanced the prestige of the wines of the region, but it was also during this period that the Loire's most famous son was born, in the town of Chinon. François Rabelais, an author, doctor, priest and celebrated bon vivant who lived and wrote in the first half of the sixteenth century, was an enthusiastic consumer of the local beverages and peppered his prose with colourful comments showing his appreciation for the wines. Even today, the Chinon winemakers say that to open a bottle of Chinon wine is to be loyal to the spirit of Rabelais, whose name inspired the local *confrérie* and who continues to be a much-quoted figure on the subject of wines, not only those from Chinon, but also many others throughout the Loire Valley.

Another important factor in the growth of the vineyard area at this time was the 1577 law passed by the Parlement de Paris, prohibiting Parisian wine merchants from sourcing their wines in the area immediately surrounding Paris. At this time, the Paris region was the major source of wines for the taverns of the capital. The bourgeois landowners producing wines of their own in the region feared the diminishing reputation of these wines by a boom in production of low-quality wines for consumption by the masses. The effect of this law, which remained in place for two centuries, was to create a surge in demand for wines from the next easiest region to reach from Paris. Notably, Orléans, with its proximity to the Loire, boomed, as did the vineyards of Touraine, which became important sources of everyday wines for the Paris market for several centuries. In 1860, the vineyard area around Orléans was recorded at 28,951 hectares, and that of Touraine exceeded 50,000 hectares (the equivalent figures today being 4,400 hectares for Touraine and under 100 for the vineyards around Orléans). Another boost for Loire wines was the effect of the 'Little Ice Age', resulting in temperatures generally cooling from the fifteenth century. It became clear that the vineyard areas north of the Loire were unsuitable for wine production (see box,

p. 5). As vineyards were abandoned in Brittany, the wines of Anjou assumed greater importance for the religious institutions of Normandy and Brittany, who now sourced their wines from there.

Orléans: more than a wine centre

While the Orléans wine business boomed, its importance as a trading port also increased, as wines from other vineyards of the Loire were unloaded from the river for onward transport to Paris. Travelling by boat from Orléans downstream to the port of Nantes took around eight days, but when wines were being transported upstream (principally from Anjou and Touraine) the journey to Orléans could take as long as 21 days, depending on the strength of the wind. By the time the wines arrived in Orléans, there was a good chance they were turning to vinegar. The Orléans merchants would taste the wines before sending them on to their clients in Paris, and any casks that were sour they kept in the city to be made into vinegar (*vin aigre* in French, meaning 'sour wine'). Orléans became the vinegar capital of France, developing the 'Orléans method' of making vinegar – a long, slow, natural fermentation in oak casks. The Orléanais gained royal recognition for their mastery of vinegar-making as long ago as 1580, and by the end of the eighteenth century there were more than 300 *vinaigriers* in the city. After the French Revolution, the industry steadily declined, as cheaper, industrialized methods of production proliferated elsewhere. Today there is only one *vinaigrier*, Martin Pouret, continuing the artisanal production of vinegar in Orléans .

Up until the nineteenth century, the Loire and its associated canals were an important means of transporting wines west towards the coast, or northwards to Paris. The development of the railways in the nineteenth century provided easy access to market but also created unwanted competition for Loire producers, whose wines were now pitted against those from elsewhere in France, notably more reliably ripe offerings from the south. But more dramatic events were in store. The arrival of the phylloxera louse in the south of France in 1863 created a temporary bubble of demand for producers further north who were not yet affected by the destructive, sap-sucking insect and could fulfil demand from the south. The Auvergne, in particular, boomed, with plantings reaching a high point of around 45,000 hectares towards the end of the century. Of course, it was only a matter of time before these vineyards were also wiped out by the all-destroying louse, never to recover. Today the vineyards of the Côtes d'Auvergne cover a mere 250 hectares.

The period following phylloxera was a new beginning. After the solution of grafting vines onto American rootstocks had been proved effective, the question then was what to plant, and where? Sites of lesser quality that had seemed attractive during the earlier boom years of the nineteenth century were disregarded as growers began to rebuild their vineyards in places with a proven history of growing quality grapes. In some places, however, the period following phylloxera saw a wholesale change. In Sancerre, for example, the previously widely planted Pinot Noir was replaced by the easier-to-grow Sauvignon Blanc, which proved to be perfectly well suited to this environment.

The chaos and profusion of unregulated, and often fraudulent, activity at the start of the twentieth century prompted the creation of the appellation contrôlée system we know today (see p. 24 for more details). This was – and remains – a way of protecting the authenticity of a product, chiefly based on its origins, but it also encompasses every aspect of a wine's production, from the type of grapes that can be planted to vine training, yields, permitted ripeness levels, vinification methods and more. The first appellations were created in 1936. In the Loire Valley, these early appellations include Muscadet Sèvre et Maine, Muscadet Coteaux de la Loire, Cabernet d'Anjou, Rosé d'Anjou, Vouvray, Sancerre and Quincy.

The twentieth century was a time of upheaval in the Loire, as elsewhere in Europe. Two world wars diminished both population and morale, but the advent of mechanization and agrochemicals promised an easier life for vignerons. Slopes that were difficult to work were abandoned in favour of flatter lands. These were often of lesser quality, but the aim was to rebuild vineyards and livelihoods and the focus was on quantity rather than quality. This period of agricultural development, known as the Green Revolution, had the same effect on viticulture as on other crops, allowing high yields to be produced year after year, often with accompanying declines in quality. This was especially noticeable in regions where wines became fashionable and production increased rapidly, as was the case in Muscadet and Saumur-Champigny in the 1980s, for example. Between 1983 and 1992 plantings doubled in the Saumur-Champigny appellation, with average yields recorded at 70 hectolitres per hectare in 1992. Yields were at the same level in Muscadet in this year, and production levels more than 60 per cent higher than they are today: in the five years between 2019 and 2023, average yields in Saumur-Champigny were 51 hectolitres per hectare and in the Muscadet appellations they were 46.

Since the latter part of the twentieth century, there has been a gradual but accelerating move towards higher quality and more sustainable wine production in the Loire Valley. Appellation regulations have been tightened to reduce permitted yields across the board, resulting in grapes with more concentrated flavours. A collective desire to reduce chemical usage is reflected in the increasing number of producers turning to organic certification and more natural winemaking processes (see p. 16). With declining wine consumption globally and a greater environmental awareness among consumers, it is perhaps inevitable that this should be the direction of travel for the Loire; nevertheless, it bodes well for the future.

The Loire's history of climate change

History suggests that Loire vineyards were developed largely during a period that was climatically highly favourable to wine growing. During this so-called 'Medieval Warm Period', which lasted from the ninth to the thirteenth century, temperatures were considerably warmer throughout Europe than they are today and vineyards were widespread in more northerly locations such as Brittany and England. Records show that before the fourteenth century the winemakers of the Orléanais were producing a highly sought-after red wine from Auvernat, otherwise known as Pinot Noir. During the same period, we see the appearance of Breton, or Cabernet Franc, which arrived in the western Loire from the south-west via the Atlantic coast.

The Medieval Warm Period was followed by the Little Ice Age, a period of generally cooling temperatures which rendered wine growing north of the Loire unsustainable. The Loire became the northern limit for wine growing in the western half of France, which further aided its development. It has remained so ever since. Loire Valley vineyards are almost all planted to the south of the river, which marks a climatic divide in France between the more northerly and more southerly regions of the country. The river moderates the climate, adding a degree or two to the temperatures and making possible production of wines which would not succeed further north. For more on the challenges of climate facing today's growers, see page 12.

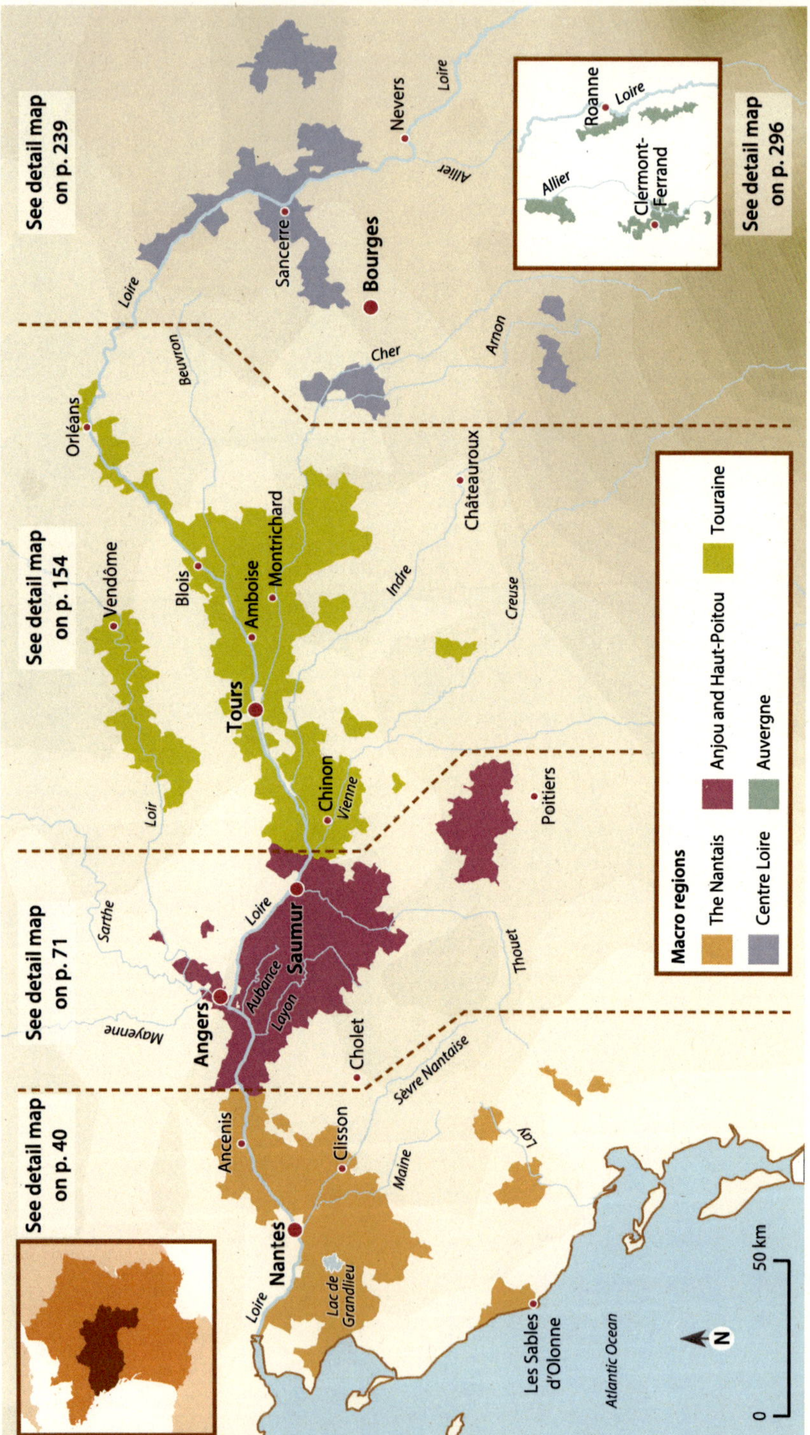

Map 1: The wine-producing regions of the Loire Valley

2

THE GROWING ENVIRONMENT

Terroir, the French wine word more laden with meaning than any other, has no direct translation, but it is usually understood to mean the natural growing environment of the vine. In France, many winemakers equate their terroir to what is going on under their vines, but the altitude, angle and aspect of the vines and the mesoclimate and microclimate can be just as important as the soil and the rocks. And then there is the human factor. A wine is not merely a product of its growing environment, but the result of hundreds of human decisions, from choosing the site and grape varieties to cultivation methods, pruning options, harvest dates, and the many choices available once grapes arrive at the winery. Compared to wine regions where grapes ripen more reliably, in the Loire there is an added emphasis on the role of the individuals responsible for tending the vines and making the wines.

THE LIE OF THE LAND

There are few wine regions as extensively connected to water as the Loire Valley, whose vineyards follow the course of France's longest river for almost 900 kilometres. The river's vast network of waterways covers a fifth of the nation's land and the vineyards extend across 15 different administrative departments: this is the most spread out wine-growing area in France. Many of the vines are clustered around the slopes of the Loire itself, but more of the Loire's appellations are influenced by its tributaries, which are often major rivers in themselves. Notable among these rivers are the Vienne (in the appellation of Chinon), the Cher (Quincy, Touraine), the Loir (Jasnières, Coteaux du Loir, Coteaux du Vendômois) and the Allier (Saint-Pourçain, Côtes d'Auvergne). In some cases the rivers define the appellations themselves: the rivers Sèvre and Maine in Muscadet, and the slopes of the Layon and the Aubance for sweet wines in Anjou, for example.

Although the Loire Valley is one of the largest wine regions in France, the vineyards are often rather discreet. Concentrated areas of vines are few; more usually they share the landscape with villages and forests, and fields of other crops, such as asparagus, sunflowers, melons, corn or wheat. The countryside of the Loire is mostly low-lying, which is no surprise given most of the vines are close to rivers. The vines tend to grow on gentle slopes that follow the contours of the rural landscape. The land rarely rises above 100 metres: a high point of the Muscadet vineyards, the Butte de la Roche, affords a marvellous view of the surrounding vineyards and the city of Nantes in the distance from a height of less than 50 metres. Exceptions to this are the hills around Sancerre, where some vines grow at altitudes approching 400 metres above sea level, and the vineyards of the Auvergne, where altitudes of 500 metres or more are common in the dramatic mountainous landscape of the former volcanoes of the Massif Central.

The soils and rocks of the Loire Valley are complex and varied, spanning a period of more than 400 million years. From the oldest rocks, to the west, formed 500 million years ago, to the 40-million-year-old flint soils around the fault of Sancerre, each appellation has its own geological profile that will influence the style and character of the wines that grow there. Often the soil and rock profiles can vary greatly over a surprisingly small area, and it is the discovery of these differences that has led to a growing trend towards producing site-specific cuvées that reflect the often unique growing conditions of each vineyard. Sancerre is a good example of this, where there has been a proliferation of wines named after specific vineyards (Les Romains, Les Demoiselles, Chêne Marchand, etc.) or soils (notably Silex, which means flint). Another development in recent years has been the application for and awarding of '*crus*', or more accurately Dénominations Géographiques Complémentaires (DGCs; see glossary) to distinguish notable sites. One successful example of this is the Muscadet *crus*. In the Muscadet Sèvre et Maine appellation (where all but one of the *crus* are located), the growers identified nine separate combinations of rocks in an area less than 15 kilometres wide. The different

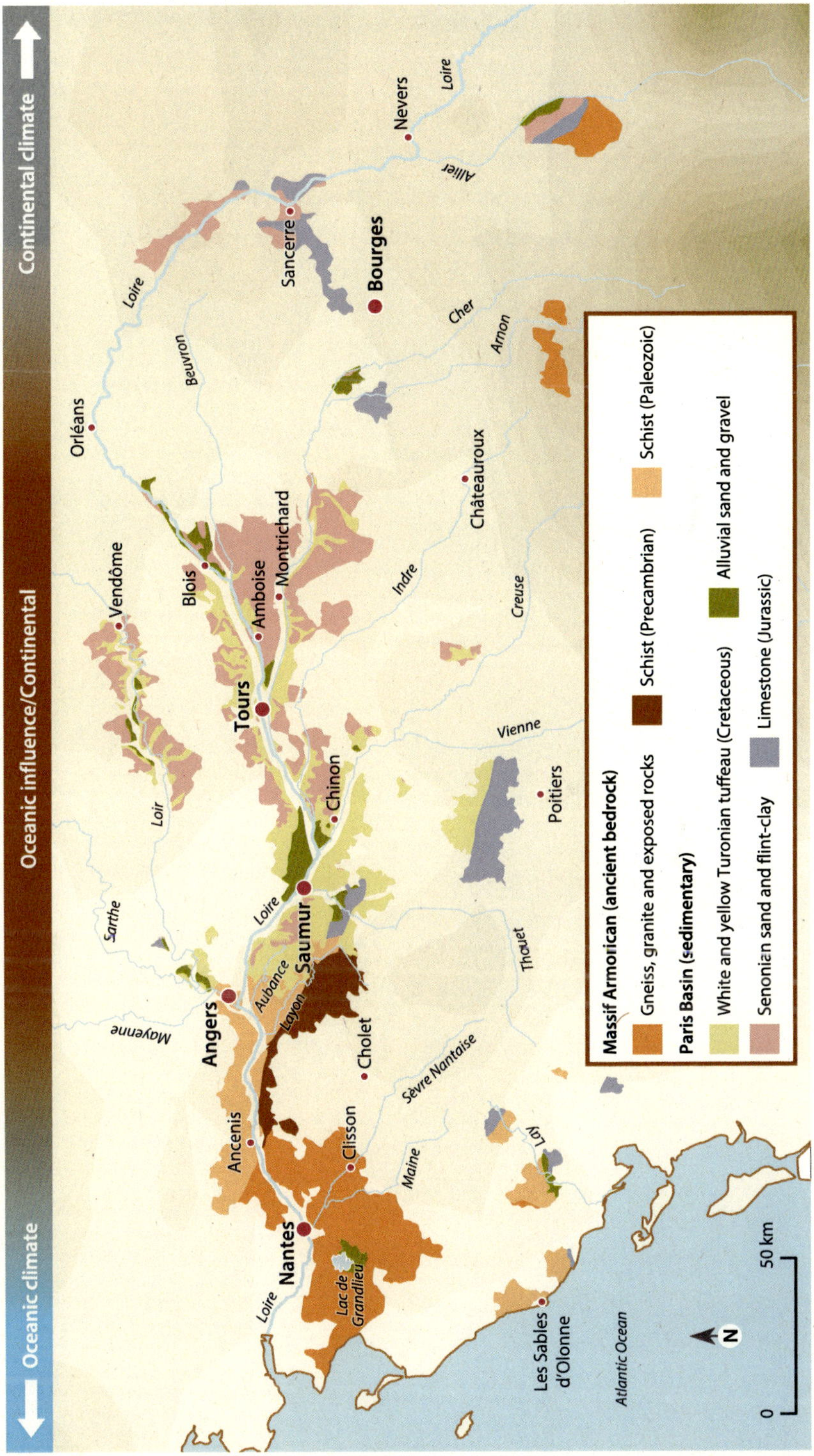

Map 2: The geology and climate of the Loire Valley

subsoils significantly affect the growing environment of the vine and the wines are radically different depending on where they are grown. The granite of the Clisson *cru*, for example, is a very warm growing medium, giving early-ripening wines with plush texture and open fruit. The neighbouring *cru* of Gorges has much cooler soils. The vines grow on gabbro, a dark, cool, clay-rich soil that is slow to warm. Here the fruit is slow to ripen. Compared to the wines of Clisson, Gorges wines are leaner, with more acid tension and minerality. A current proposal is to create a similar set of DGCs for the Saumur appellation for white wines.

Despite the geological complexity, the rocks that lie under the vineyards can be divided into three main categories. In the far west, the vineyards of the Nantais and the western part of Anjou (the area west of the city of Angers, known as Anjou noir, or black Anjou), are the ancient igneous and metamorphic rocks of the Armorican Massif. This is the bedrock of all of north-west France, the source of Brittany's rugged coastline and the complexity of the Muscadet vineyards. These acidic rocks include granites and schists, gneiss, gabbro, slate, orthogneiss, serpentinite and amphibolite, among others.

Loire appellations on volcanic and metamorphic rocks

Armorican Massif

- Muscadet appellations
- Gros Plant du Pays Nantais
- Coteaux d'Ancenis
- Fiefs Vendéens
- Anjou noir appellations: Anjou, Anjou-Villages, Anjou Brissac, Anjou-Coteaux de la Loire, Cabernet d'Anjou, Rosé d'Anjou, Coteaux du Layon, Coteaux du Layon Premier Cru Chaume, Quarts de Chaume Grand Cru, Bonnezeaux, Coteaux de l'Aubance, Savennières, Savennières-Roche aux Moines, Coulée de Serrant

Massif Central

- Saint-Pourçain
- Côtes d'Auvergne
- Côte Roannaise
- Côtes du Forez

Anjou noir and blanc: black schist and white tuffeau in the massive walls of the Château of Angers

To the east of Angers (the Anjou blanc, or white Anjou), the land comprises sedimentary rocks that date mostly from the tertiary era. These are the alkaline limestones and marls formed from the compaction of sediments in the Paris Basin, a former sea that covered an area from the east of Angers through the whole of the Touraine vineyard, stretching as far north as Paris and extending east to the central vineyards in and around Sancerre (this same sea also covered the vineyards of Burgundy and Champagne). Anjou appellations in the Paris Basin include Saumur and Saumur-Champigny, Saumur-Puy-Notre-Dame and Coteaux de Saumur. All of the appellations in the Touraine and the Centre Loire region are within the Paris Basin as well.

The third category is the rocks of the vineyards of the upper Loire, which were formed from the Massif Central, a series of former volcanoes that dominates the landscape today. The winegrowers of the appellations in this region – Côtes du Forez, Côtes d'Auvergne, Saint-Pourçain and Côte Roannaise – have organized themselves under the marketing banner of Loire Volcanique, reflecting the influence of this landscape on their wines. Like the Armorican Massif further north, the terrain here is a complex mosaic of rocks, mainly a mixture of ancient and newer volcanic origins that includes basalt, granite, schists and gneiss, as well as areas of clay-limestone and alluvial terraces.

CLIMATE AND WEATHER

At a latitude of 47 degrees north, the Loire Valley is one of the most northerly vineyard areas in Europe and this location is the most important influence on its climate. Within France, only Champagne and Alsace are further north (each with a more continental climate, being further inland and to the east). This latitude was considered to be at the marginal limits of successful viticulture in the middle of the twentieth century, but a rise in average temperatures of 1.8°C between 1960 and 2010 means that the Loire climate has become more temperate. This is a trend that has continued in recent decades, accompanied by higher concentrations of sugar in grapes, lower levels of acidity in wines and earlier harvest dates. An ongoing study shows average temperatures to be rising almost every year, with a recorded increase of 3°C in the last 30 years. None of this is to say the climate is reliably warm, however, and there are annual challenges for winegrowers to face. Many now say this unpredictability is the most difficult aspect of wine growing for them to manage, although the issues are often different here compared to warmer parts of the planet.

The Loire has, in many respects, experienced positive effects from warmer vintages. This far north additional heat and sun during the growing season can be beneficial for ripening wine grapes, although leading Pouilly Fumé producer Louis-Benjamin Dagueneau was quoted as saying that since the warm 2018 vintage, 'we are reaching the limits of the

Sauvignon grape variety'. For reds, the heat has been less problematic, the additional degrees of temperature especially helpful for subduing the green, leafy characters to which Cabernet Franc is prone. The Loire has seen its best run of good harvests in recent years: 2018, 2019, 2020 and 2022 were previously once-a-decade vintages, but are now almost expected. It is in expectation that difficulties lie, however, because every year seems to bring new challenges. Time and again winemakers comment that every year is different now: the new normal is the unknown, although the effects of spring frosts are now so significant that most wine growers have an annual plan to cope with this risk (see box overleaf).

As I write this, in summer 2024, it has rained more than anyone could have imagined. Wine growers despair that there is still no reliable way to predict the weather more than a couple of days in advance. In 2023, it was also cool and rainy; in 2022 there was heat and drought, and 2021 was the year where, in the words of one winemaker, 'we had everything but the locusts'. Speak to any winemaker about vintages and they will remember in vivid detail how many days of frost they suffered, how much crop they lost, the amount of rain (or worse, hail) that fell in a particular week, or how many days of drought before it rained, or – just occasionally – the bountiful harvest of perfectly ripe grapes. Every appellation will have its own challenges and, other than the general trends, there can be a lot of variation in growing conditions between them. Over the years, the vignerons have adapted to the climate, but when it is perceived to be changing every year, this becomes more difficult. After 2018, 2019 and 2020, it seemed as if warm vintages were the new normal. And then came 2021. The talented winemaker is the one who can modify their approach to suit the conditions of each vintage. And the best growers somehow manage to walk this tightrope successfully, year in and year out.

Within the regions there are significant differences in climate that affect the grapes which can be grown and the styles of wine made. The vineyards of the Nantais, towards the estuary of the Loire, have a strongly oceanic climate. Some of the vineyards of the Fiefs Vendéens even grow within sight of the Atlantic coast. Notably cooler than other regions, and with higher rainfall, the Nantais is most suitable for growing white wines, which make up most of the production (and are mostly Muscadet). Moving eastwards, Anjou has a temperate oceanic climate; the power of the ocean is still felt, but to a diminishing degree. Here, as elsewhere, other factors such as proximity to the river (an influence on temperature, airflow and humidity), or orientation of slope, become more important. Beyond Anjou, the climate becomes steadily more continental towards Touraine, and the temperatures warmer. Anjou is the most diverse of the Loire's regions for wine styles at all quality levels. From Saumur eastwards is the nexus of red wine production in the Loire: the vineyards of Saumur-Champigny, Chinon, Bourgueil and Saint-Nicolas-de-Bourgueil. The Centre and Upper Loire have more continental climates, with hotter summers and cooler winters.

Growing season sunshine hours, average temperatures and rainfall for the main Loire regions (source: InterLoire)

1 April–30 October	Nantais	Anjou and Saumur	Touraine	Centre Loire
Sunshine hours (cumulative)	1,570	1,420	1,395	1,372
Temperature (daily average)	14.4°C	15.4°C	15.3°C	15.3°C
Precipitation (cumulative)	406 mm	352 mm	376 mm	527 mm

Managing frost

One of the effects of climate change in the Loire Valley is that winters are generally warmer, so the growing season starts earlier. This means that early vine growth is at risk from spring frosts which, despite the often warmer growing seasons, have not abated in either frequency or severity. Some appellations and varieties are more susceptible than others, but 2016, 2017, 2019 and, especially, 2021 saw devastating crop losses in the Loire vineyards, sometimes involving destruction of the entire year's crop, motivating growers to search for solutions to save their grapes.

A few different options have emerged. One is to delay winter pruning in order to retard the spring growth of the vine. Another popular choice is to prune twice. The first time the fruiting canes are pruned longer than is needed. Since the vine's growth begins in buds at the furthest tip of the cane, any frost damage will affect this part of the plant first. After frost or risk of frost has passed, the vineyard workers make a second pass in the vineyard for a final pruning to the desired number of buds. This increases costs, but it is money well spent if it saves the crop.

Other solutions require more substantial investment. Tall frost fans are now a common sight in Loire vineyards, although at 65,000 euros each, this is not an option everyone can afford. In some places (Quincy and Chinon, for example), growers have come together to invest collectively in protecting large swathes of land for the benefit of everyone. Chinon producer Mathieu Baudry recalls how after the frosts of 2016 the growers accelerated their frost protection, collectively installing 50 frost towers.

In some places it is possible to use water aspersion, an effective anti-frost treatment, but this requires a substantial water supply, as the vines must be sprayed throughout the night to ensure protection. Despite the ubiquity of rivers here, there is a shortage of freely available water sources for this type of activity.

The most fail-safe, but also the most expensive, option that has become available recently is to install heated electrical wires in the vines. This raises the temperature in the growing zone just above the critical level where damage occurs. It is a method used by high-end producers such as Clos Rougeard, Domaine Belargus and Thibaud Boudignon. The only constraint of this method, according to Thibaud Boudignon, is that it is not suitable to protect very old vines because their gnarled and twisted shapes impede installation.

THE HUMAN FACTOR

Key to quality at every stage of wine growing are the men and women who dedicate their time and energies to this profession. This is a producer-focused book because ultimately, it is people who make the difference, and in whose hands success or failure lies. The best vineyard in the world will not produce great wine all on its own (despite the protestations of those winemakers who claim to 'do nothing'). Although the place where the grapes grow could be considered the soul of a wine, its character is very much determined by human intervention.

There have been significant changes in approaches to wine growing in the last two decades. A move towards more conscious production has taken the form of more controlled yields and greater recognition for the environment and biodiversity. Some appellations (such as Saumur-Champigny) now require growers to include green cover between their vines, and growers across the region are planting trees and hedges, or creating wildlife 'corridors' to increase biodiversity.

Winemaking styles have also changed and, although not universal, there is a noticeable shift towards making wines that are more suited to the region – *vins ligériens* (see box, p. 16) – as well as to the modern palate.

Towards the end of the twentieth century, there was a trend to produce red wines with heavy oak influence, emulating the style of France's more famous growing region for Cabernet Franc, Bordeaux. But growing conditions in the Loire are very different from those further south, and the wines were often distinguished by rather weedy fruit accompanied by astringent tannins from extended maceration and pumping over during winemaking, in addition to powerful aromas and tannins from lengthy oak ageing. But in the last twenty years, growers have turned away from trying to emulate Bordeaux and have instead focused more on playing to their natural strength of being able to produce naturally lighter, fresher styles. Extractions have been reined in: more producers refer to 'infusion' now, and not just for Cabernet Franc, but across all red wine varieties. Helped by a string of warm vintages, growers have realized that easing off on the extraction allows for more transparent wines that reveal the true nature of the fruit and reflect more accurately the place in which they were grown.

Le style ligérien

A word you hear quite often in the Loire Valley is *ligérien*, which literally means 'coming from the Loire'. But when a winemaker describes a Loire wine as '*un vin ligérien*', they are not simply stating a self-evident truth, but talking about a particular style of wine. A *ligérien* wine is one that is refreshing and light on its feet, balanced, with no excess of alcohol or oak, other obvious winemaking manipulations, or overripe grapes. Flavours will be at the crunchy fruit end of the spectrum, rather than tropical or confit. *Ligérien* is not a descriptor of quality, although it is invariably used in a positive sense. Essentially, it is a recognition of a style that is classically, and uniquely, of the Loire.

A NATURAL APPROACH

From grape growing to winemaking, things are a lot more natural than they used to be in the Loire Valley. The number of producers embracing an increasingly chemical-free approach is growing every year. France, Spain and Italy lead the world in terms of hectares certified organic, and within France the Loire is one of the most important regions for organic viticulture. In the decade from 2011 to 2021 the area of vineyards farmed organically here rose by nearly 300 per cent and now accounts for 23 per cent of the total land area. This compares to the French national average of

14 per cent. It is often the organic growers who are making the most interesting wines, too. Of the profiles included in this book, more than 80 per cent are certified organic. It was not a conscious decision to seek out organic producers for this book, but very often these are the most interesting growers. In their search to make the best decisions for their vines, for the environment and for their wines, more and more see organic viticulture as a natural choice. It's no coincidence that these wines are often the best.

As well as having a relatively high proportion of organic and biodynamic growers, the Loire is also a centre for the natural wine scene. The word 'natural' has no fixed definition, but the most common (and commonly debated – see box, p. 18) feature of these wines is the absence of the sulphites that are present in almost all other wines. Derivatives of sulphur are used at several stages of winemaking, from grape reception through to bottling. Sulphur is also a key tool in combating fungal diseases in vines, especially in organic viticulture. In winemaking, sulphur dioxide is valued for its antioxidant, antimicrobial and antiseptic properties – all helpful when making a perishable product for human consumption. The downside of sulphur, though, is that it can negatively impact the sensorial aspects of wine if used to excess (and a small percentage of people are allergic to sulphites). Excessive sulphur use would certainly have been a fair accusation to level at some of the Loire wines in the latter half of the last century. But attitudes to sulphur have taken a 180-degree turn. Now, the frequently heard intention from winemakers is to use as little sulphur as possible, or even to avoid it altogether. Choosing not to use sulphur at any stage of winemaking or bottling exposes wine to risks most winemakers are unwilling to take, such as oxidation, or risk of infection from bacteria or other spoilage organisms. On the other hand, the ideal of unadulterated, 'pure and natural' wine has undeniable appeal, because anything added to wine is likely to have some effect, however slight, on its taste.

While total exclusion of sulphur from the winemaking process is relatively rare, many winemakers have drastically reduced the amount that they use. This is a welcome directional change. The question nowadays, however, is: how low can you go and still have a wine that is protected? There is no straightforward answer to this because it depends on a number of things, such as the style of the wine, its chemical composition (which will depend on the vintage, among other things), whether it needs to survive lengthy transportation and, crucially, when it is intended to be drunk, because the longer a sulphur-free wine remains in bottle, the greater the risks of spoilage.

The natural wine debate

Who wouldn't want their wine to be as natural as possible? This is certainly the view of the Syndicat de Défense des Vins Naturel, whose charter for *Vin Méthode Nature* has been adopted by a number of vignerons who prefer to eliminate sulphur from their wines altogether. In order to qualify for sticking the mildly subversive-looking handwritten logo *Vin Méthode Nature* on a wine, it must be sourced from organically certified vineyards and made from manually harvested grapes, have no oenological additions, no filtration, and no sulphites added before or during fermentation. For those who can't quite stomach zero sulphites in their wines, there is a second logo for wines with under 30 milligrams per litre of sulphites ('*< 30 mg/L de sulfites*').

So where lies the problem? Essentially, it's the wines. Failing to use any sulphur in wine is a risky business and, while there are some truly excellent sulphur-free wines, many more are controversial. Oxidation is commonplace in these wines, and evidence of *Brettanomyces* is far from rare. And then there is the uniquely unpleasant 'mouse taint'. And further 'unusual' flavours that come from the more natural approach. Interestingly, characteristics that most traditional winemakers and drinkers consider aberrant are actually welcomed by natural wine enthusiasts, including some wine professionals. One winemaker showing me his son's sulphur-free wine said to me, 'of course, we are too old to appreciate this kind of wine', because it's true that there is a greater acceptance for the flavours of 'natty wine' (it even sounds cooler than boring old 'wine') amongst younger drinkers. So, there is a generational divide, but also a quasi-religious dogma about natural wines, of the 'you're either for us or against us' variety.

Some say the movement is about the cult of the individual and the erasure of terroir, meaning that by producing wine 'naturally', ironically the essence of the terroir is actually lost, because one oxidized wine tastes very much like another, in the same way that Bretty wines – dominated by aromas of farmyards and old leather – tend to taste very similar.

But there is also a middle ground. Many people are tolerant of these characteristics in small measures. If a wine is not intended for cellaring, why not drink it young and fresh and free from any additives (before any possible deviant flavours have had the chance to develop)? And then there are the sulphur-free wines that transcend others with their purity of fruit. It's a debate that is far from over.

RISING DEMAND

In contrast with other regions, the latest figures for vineyard transactions show that the Loire is bucking the trend across France for declining land prices. In the predominantly red wine regions of Bordeaux and Côtes du Rhône the situation has reached crisis point. With declining demand for red wines and fears of an unsustainably warming climate in the south, buyers are looking north. The Loire offers the flexibility of a cooler climate that is ideal for the production of more refreshing wine styles of any hue. These wines are becoming both harder to find and more highly demanded by consumers. In 2023, land prices in Sancerre were booming, at an average price of 260,000 euros per hectare. Prices are also on the rise in less prestigious Loire appellations, up 17 per cent in Touraine and 19 per cent in Saumur. There is evidence of a renewed interest in land in the Muscadet *crus*, with land prices increasing here as well. Despite these rises, the Loire has the added attraction of remaining affordable in most places. For example, the 17 per cent increase in the average per hectare price of Touraine land raises the price to only 14,000 euros.

The affordability of the Loire has attracted an influx of new talent. Young and new vignerons in particular have been able to set up small domaines across the region, encouraged by the dynamic natural and environmentally friendly wine scene, an enticing palette of grape varieties and styles to choose from and, on a more practical level, simply by being able to find suitable properties available to buy, which in some regions (Burgundy, for example) simply do not exist.

3

WINE STYLES, APPELLATIONS AND GRAPE VARIETIES

WINE STYLES OF THE LOIRE

The Loire has a more diverse range of wines than any other region in France, made from a broad range of grapes. But for all their variety and the distance that separates the many appellations, Loire wines have much in common. For a start, they share an inherent drinkability: most can be drunk soon after bottling and most are made to be enjoyed young. Another key feature of Loire wines is that they are invariably refreshing.

Most Loire wines are made from a single grape variety. One might think this would make them easier to understand from a consumer's perspective. In some cases this is true but, as with most French wines, the name of the grape variety rarely features on the label of the bottle. Below is a guide to the appellations where the different varieties grow.

White grapes thrive in the cool temperatures of the Loire Valley, so it is no surprise that whites are the most commonly encountered wine style, at 44 per cent of production. Sauvignon Blanc, Chenin Blanc and Melon are the major grape varieties, and the top Loire examples are the finest wines made from these grapes anywhere in the world.

Chenin Blanc is the most emblematic white grape variety of the Loire, where it displays its versatility in producing a wide variety of styles at every level of quality, from sparkling to botrytized and all levels of sweetness in between. Chenin is a wonderful variety for transmitting terroir, and it

The Loire at a glance

- **Number of appellations**: 53
- **Area under vine**: 56,000 hectares
- **Area classified AOC**: 45,600 hectares (81 per cent of total)
- **Styles (2023)**: 44 per cent white, 22 per cent rosé, 18 per cent red, 16 per cent sparkling
- **Major grape varieties**: Cabernet Franc, Sauvignon Blanc, Chenin Blanc, Melon
- **Number of grape varieties permitted for IGP (regional) wine**: 33
- **Number of individual producers (2021)**: 3,700, plus 21 cooperatives and 450 négociants
- **Production (2023)**: 320 million bottles
- **Markets**: 74 per cent domestic, 26 per cent export
- **Top three export markets**: USA, UK, Germany

finds many different expressions in the Loire. It is the sole grape of the Loire's only *grand cru* and *premier cru* appellations, Quarts de Chaume Grand Cru and Coteaux du Layon Premier Cru Chaume, as well as the only permitted grape for all of the other sweet wine appellations of the Layon Valley and Montlouis-sur-Loire. In practice, it is almost always the sole grape for the wines of Vouvray as well. Much of the Loire's Chenin harvest is destined for sparkling wine, whether from Vouvray, Saumur or Montlouis, or as a key component in Crémant de Loire wines.

The crisp, mineral Sauvignons from Sancerre are by far the Loire's most successfully exported wines, with almost two-thirds of production leaving the country every year. As well as Sancerre's immediate neighbour, Pouilly Fumé, Sauvignon is also the mainstay of white production in the various Centre Loire and Touraine appellations.

Melon, an old variety that has found its home in the vineyards of Muscadet, where it is grown almost exclusively, produces drinkable and food-friendly wines. But in some locations, and in the right hands, these understated wines are transformed into some of France's finest. Melon is slowly being recognized as an excellent vehicle for interpreting the complex terroirs of Muscadet. With careful handling and extended ageing on lees, these are unique wines that display some of the best qualities of the Loire: subtlety, restraint and balance. Other white varieties ubiquitous elsewhere in the wine-growing world – Chardonnay, for example – make fleeting appearances in the Loire, either confined to a single appellation or, more usually, as part of a blend.

Rosé is the second most significant wine style in the Loire, at 22 per cent of production. Most is inexpensive and off-dry or semi-sweet. While the Loire's professional body boasts that it is the only region in the world to promote this sweeter style of rosé, it does little to elevate the reputation of the Loire as a source of quality wine. Only 5 per cent of the sweetest and most popular style, Cabernet d'Anjou, ever leaves France. Drier rosés have profited from the international fashion for pink wines and have made greater inroads into overseas markets. There are some attractive Cabernet Franc rosés to be found, from the good red wine producers of Chinon and Bourgeuil in particular.

Sparkling wine (pink, white, or even, occasionally, red) is also an important and growing wine style in the Loire, especially in Saumur and Vouvray, where the extensive tuffeau cellars provide perfect conditions for these wines to age. Most is made with two fermentations in the traditional (Champagne) method and ages for nine months to a year before release. On a much smaller scale, a number of producers make sparkling using the so-called 'ancestral' or *pétillant naturel* method, or some version of it. This is a single fermentation made using riper grapes with high sugar levels. The fermentation starts in the usual way, and is then arrested by chilling to stop the yeasts' activity, before transfer to bottle to finish the fermentation using the sugar remaining in the must. The process gives a naturally sparkling wine, sometimes with lower pressure than the traditional method. In practice, it is common for producers who make this style to add extra yeast when they transfer the fermenting must to bottle in order to restart the fermentation efficiently. This is not permitted in Montlouis, however, where producers have their own appellation for this style of wine, called *pétillant originel* (see p. 173).

Red wines account for 18 per cent of production. The major area for quality reds is centred around the triangle of Saumur, Chinon and Bourgueil, which offer different expressions of Cabernet Franc, the Loire's most planted grape. Further east, in the vineyards of Sancerre and the other vineyards of the Centre, proximity to Burgundy has shaped the choice of varieties over the years. Pinot Noir is the preferred grape here; further south (not so far from Beaujolais), Gamay thrives in the acid soils of the Massif Central.

Developing new grape varieties for the future

One of the major challenges of grape growing in the Loire is mildew, which requires significant application of chemical plant protection products. Grape vines are especially sensitive to fungal diseases, and viticulture is a significant user of agrochemicals. The French agricultural and environmental research institute, INRAE, estimates that French vineyards account for 20 per cent of the country's agricultural chemical use, despite only taking up 3 per cent of arable land. The European Union has a radical plan to reduce chemical use in agriculture. One way that researchers are trying to accommodate this is by introducing new varieties that are disease resistant, explains Sam Schrock, of the French Vine and Wine Institute (IFV). The ResDur regional breeding programmes have been underway for 20 years throughout France in an attempt to solve the twin problems of climate change and excessive use of chemicals in vineyards. The aim is to develop varieties that are genetically resistant to downy mildew and powdery mildew, but also suitable for producing high-quality wine. It is a lengthy process, involving extensive breeding and testing. At the IFV offices just outside Angers are dozens of micro-vinifications of these new varieties, bearing such novel names as Floréal, Voltis, Artaban, Sirano, Coliris and Lilaro. The appeal is evident: they can reduce mildew by up to 99 per cent, compared to untreated traditional varieties that would suffer almost 100 per cent loss without some form of chemical treatment. While some of these new varieties are officially approved for inclusion in the IGP Val de Loire appellation already, current work is focusing on improving the flavour profile further by crossing with existing local *Vitis vinifera* varieties. 'The aim is to create a variety that is as close as possible to the vinifera varieties we know and love,' says Sam. In Angers, this work is being done on Chenin Blanc, Sauvignon Blanc and Melon. In October 2023 AOC Touraine became the first appellation in the Loire to request access to Floréal for experimentation for its suitability in the region in view of the fact that Touraine Sauvignon Blanc has seen a decline in acidity levels and a corresponding rise in sugar levels in the last 30 years. Floréal was selected due to its similarity to Sauvignon in terms of aroma, as well as its resistance to mildew.

Climate change has had some effect on the styles of wine produced, giving richer fruit and lower levels of acidity in warmer vintages. (It has also brought other challenges that have prompted research into alternatives for the future; see box, above.) Some styles of wine – particularly whites – have on occasion suffered from unbalanced alcohol levels in recent years, while others – reds from Cabernet Franc, for example – have certainly benefited from the additional ripeness of warmer growing seasons. And although acid levels may be lower, the Loire as a region is never short of acidity compared to other places. It is also reassuring that one of the Loire's major calling cards, Chenin Blanc, thrives in the hotter, drier climate of South Africa, where it also retains its acidity despite a warmer climate.

THE APPELLATION SYSTEM IN THE LOIRE VALLEY

France invented wine classification and, as a country famed for its paperwork, it has made the business of wine law very comprehensive indeed. The Institut National des Appellations d'Origine (hereafter known by its more familiar acronym INAO), was set up at the start of the twentieth century, following the wine industry's limping recovery from the phylloxera catastrophe at the end of the nineteenth century. With the destruction of the country's entire vineyard in a couple of decades, all notions of norms and tradition were abandoned; grape growing and winemaking became something of a free-for-all. Fraud was rife. There was generally no way of knowing what was in your glass, nor where it came from, nor whether it had been adulterated along the way.

The INAO set about changing all this by introducing a system of rules and regulations designed to protect legitimate producers from unscrupulous imitators. The rules designated boundaries for production, stating which grape varieties could be planted, how they should be grown, the size of the harvest and winemaking practices that needed to be followed. The first appellations were granted in 1936, among them several of the Loire wines we know today. These include the AOCs Muscadet Sèvre et Maine, Muscadet Coteaux de la Loire, Sancerre, Vouvray, Quincy, Rosé d'Anjou and Cabernet d'Anjou.

Since its creation, the INAO has approved hundreds of appellations and created many thousands of rules to be upheld by those wishing to claim appellation status. There is a rule for everything, for every appellation, from the percentages of varieties permitted in a blend right down to the space allowed between plants and the maximum percentage of missing vines allowed per hectare. They've thought of everything. There are 53 appellations in the Loire Valley, and sub-appellations (or Dénominations Géographiques Complémentaires – see glossary) within several of these.

The INAO has also created IGPs, or Indications Géographiques Protégées (protected geographical indications). These classifications, formerly known as Vins de Pays, or country wines, were created in 2009. They have more relaxed rules, notably regarding the choice of grape varieties that can be used, and there is a more generous maximum permitted yield of 90 hectolitres per hectare. The largest IGP of the Loire Valley is IGP Val de Loire, which includes land in all the 14 departments of the Loire Valley and allows 33 different varieties to be used, singly or in blends. In practice, 90 per cent are varietal wines. As Val de Loire is such a broad area, the name may be completed with a closer geographical indication. These are: Allier, Cher, Indre, Indre-et-Loire, Loire-et-Cher, Loire-Atlantique, Loiret, Maine-et-Loire, Nièvre, Sarthe, Vendée, Vienne, Marches de Bretagne, and Pays de Retz. In the vineyards of the Centre are the small IGPs across the river from Sancerre, Côtes de la Charité and Coteaux de Tannay, as well as Coteaux du Cher et de l'Arnon. In the Auvergne are IGPs Puy de Dôme and Urfé. In 2023 IGP production accounted for 14 per cent of the Loire's volume.

While some producers take advantage of the higher yields under IGP to make high-volume wines of limited interest, others use the more flexible rules to make wines that are not precisely within appellation rules, such as a single varietal wine in an appellation that requires all wines to be blends. Many of the latter group, however, prefer to declassify their wine altogether to the most basic appellation of Vin de France, a category that is a popular choice among young and 'natural' winemakers. Vins de France make up a significant 17 per cent of the Loire's output.

The mention of the word *cru* conjures up images of the fine wines of Bordeaux and Burgundy. In the Loire, until quite recently there was no official mention of the word anywhere, even if the best sweet wines of the Layon Valley had long been known for their high quality and, in texts written before the INAO was created, the best wines were referred to as '*crus*'. In 2011 this changed when the Loire received its first – and still only – *grand cru* wine appellation, Quarts de Chaume. This is a designation that is permitted only for sweet wines. It sits on the same hill as the appellation Chaume, now classified as Coteaux du Layon

Premier Cru Chaume. This is the Loire's only *premier cru* appellation and this, too, is for sweet wines only. The growers in these two appellations have been lobbying the INAO to create another *premier cru*, known as Ronceray, for the dry wines that come from these two appellations. This has yet to be approved, although the name is now commonplace on bottles of dry wine from these terroirs.

Growers in other parts of the Layon Valley would like their sweet wine appellations to allow inclusion of dry wines to reflect the fact that most producers now make sweet and dry styles from the same terroirs. There is clearly a logic to this. But the INAO favours keeping things simple for the consumer, so its current position is that these growers need to apply for a separate classification for dry wines, with a different name from the sweet wine appellation to avoid confusion for the consumer who won't know if an appellation's wine will be sweet or dry (although arguably a raft of new names will in itself be quite confusing for the consumer).

In Anjou and Saumur, growers are working towards achieving *cru* status for specified existing dry white wine sites, but if INAO's progress towards granting the Muscadet *crus* is anything to go by, this could take a while. The first three of the ten Muscadet *crus* were granted in 2011 after a decade of work by growers to show that the wines were worthy of special status. As François Robin, head of communications for the wine growers of Muscadet, points out, 'The aim of INAO is not to create something new, but to recognize the specific conditions of a place and of a wine.' In other words, to achieve recognition for classification, you need to show that something already exists. In this context, the word *cru* means Dénomination Géographique Complémentaire, in other words a specific geographical mention added to the existing appellation with the hope that later recognition will follow in the form of a separate appellation in due course. And then, perhaps *premier cru* status after that – although it will probably be the current growers' grandchildren applying for that.

GRAPE VARIETIES OF THE LOIRE

Listed below each variety are the appellations where the variety is permitted. For appellations in bold letters, growers may only use this variety in their wines. (If an appellation includes wines of different colours, the bold text refers to the colour of wines usually made from that grape.)

The main white grapes

Sauvignon Blanc – around 10,000 hectares

Cheverny, **Coteaux du Giennois**, Fiefs Vendéens, Haut-Poitou, **Menetou-Salon**, **Pouilly Fumé**, **Quincy**, **Reuilly**, Saint-Pourçain, **Sancerre**, Saumur Mousseux, Touraine, **Touraine-Chenonceaux**, Touraine-Mesland, **Touraine Oisly**, Valençay

Sauvignon Blanc is the most planted white wine grape in the Loire Valley and the region accounts for nearly a third of all France's plantings of the grape. Most likely a native of the Centre Loire region, there were known to be plantings of Sauvignon in the Sancerre and Pouilly areas from 1783. Today Sauvignon is still the lynchpin of the Central appellations, where it is the only permitted grape for white wines. Sauvignons from Sancerre and Pouilly Fumé are typically fermented in stainless steel and bottled when young and fresh for export. They offer a less obviously pungent and fruit-forward style compared to New Zealand Sauvignons, erring on the side of cool restraint, as indeed most Loire wines do. Flavours of grass, nettles and elderflower are common. On the best sites of Kimmeridgian marls, stony clay-limestone soils or flinty clays, the wines can show savoury, mineral and complex flavours and, the winemakers hope, not taste like Sauvignon Blanc at all. These terroir-expressive, more complex wines are frequently aged on lees for a time in older oak and merit at least a few years in bottle to show at their best. Other Sauvignons from the central vineyards of Menetou-Salon, Quincy, Reuilly and Coteaux du Giennois offer wines in a similar style. What they lack in precision and finesse they make up for in value for money. Menetou-Salon, being geographically and geologically closest to Sancerre, is the most similar in style.

Touraine is also an important source of good-value wines from Sauvignon, where it typically produces light, floral and airy varietal styles for early consumption. The sub-appellations (or Dénominations Géographiques Complémentaires) provide variations in the grape's expression, and are usually of higher quality than the basic appellation. The site-specific Touraine-Chenonceaux wines have added richness from lees ageing, while Touraine-Oisly wines grow on sandy soils that give particularly rich, slightly weightier wines with flavours that can verge on tropical.

Sauvignon is a common blending partner in Touraine with Sauvignon Gris. In Cheverny and Valençay it is found in wines as a majority grape with Sauvignon Gris, Chardonnay, Chenin or Menu Pineau, and it sometimes plays a supporting role to Chenin in the whites of the Fiefs Vendéens.

Chenin Blanc – around 9,500 hectares

Anjou, **Anjou-Coteaux de la Loire**, **Bonnezeaux**, Cheverny, **Chinon Blanc**, **Coteaux de l'Aubance**, **Coteaux du Layon**, **Coteaux du Layon Premier Cru Chaume**, **Coteaux du Layon 'Villages'**, **Coteaux du Loir (white)**, **Coteaux de Saumur**, Coteaux du Vendômois, **Coulée de Serrant**, Crémant de Loire, Fiefs Vendéens, **Jasnières**, **Montlouis-sur-Loire**, **Quarts de Chaume Grand Cru**, **Saumur**, Saumur Mousseux, **Savennières**, **Savennières Roche aux Moines**, **Touraine Amboise**, **Touraine Azay-le-Rideau**, Touraine-Mesland, Vouvray

Chenin is the Loire's still-underrated trump card for white wines, being both its most versatile and arguably its most noble variety. It has been cultivated in the Loire Valley for more than five hundred years and was historically known as Pineau de la Loire or Plant d'Anjou in recognition of its most probable origin.

The Loire has more than 90 per cent of French plantings of the grape and it is present in most of the western Loire's designated appellations. Chenin can produce every style of white wine. Its reliably high acidity provides backbone and mouthwatering lift to sparkling and sweet whites alike, while the fully ripe, dry styles can compete with the best Chardonnays for complexity, weight, richness and longevity. Chenin is a variety that offers the drinker wines with a real sense of place, from the bone-dry, linear styles of Saumur to the honeyed, lanolin breadth of Vouvray or the mineral-laden, chiselled complexity of Savennières and Anjou blanc. When it comes to sweet wines Chenin excels, achieving high ripeness levels while losing none of its scintillating acidity. It has a naturally honeyed character and a predisposition to botrytis, making it an ideal candidate to produce highly complex sweet wines that can age beautifully. Chenin is the sole permitted variety in all of the great sweet wines of the Loire, as well as Saumur and Montlouis-sur-Loire (the wines of Vouvray are also usually 100 per cent Chenin, but may contain up to 5 per cent Menu Pineau). It is also by far the most important variety for production of sparkling wines, comprising at least 70 per cent of any Anjou sparkling wine and 80 per cent of Anjou blanc still wines. The rich diversity of soils, microclimates and winemaking personalities create an extravagant variety of Chenin wines in the Loire Valley.

When young, the grape's always-brisk acidity gives crisply green, crunchy wines, with fruit profiles of apples and lemons, quince, white peach and white flowers. As the wines age, more exotic notes may appear: yellow peach, mango and honey are common after three or four years. For decade-old wines, expect more honey, candied fruits, marmalade, apricots and figs, especially for riper, botrytized wines. Winemaking choices such as lees ageing, malolactic fermentation and oak can add biscuity, buttery, creamy characters, and flavours of vanilla, cinnamon or cedar.

Chenin is not an easy grape to grow. Bud break arrives early, putting vines at regular risk of spring frosts, and it is susceptible to diseases such as mildew and rot. Chenin has many clones and Etienne Neethling, a researcher working for the French national research institute, INRAE, in Angers, notes that there can be three weeks' difference in ripening time between different clones of Chenin Blanc. He sees a future where more clones will be used to give even greater complexity and diversity, plus insurance against the unknown climate events of the future. In the town of Montreuil-Bellay there is a Chenin Blanc conservatoire growing 433 clones of the variety. Of these, 14 are certified for current use, although in practice only two are used for the vast majority of new plantings, 220 and 880. Some growers (Thibaud Boudignon in Savennières, for example) prefer to create diversity by using a massal selection for their new plantings.

Melon – around 8,500 hectares

Muscadet, **Muscadet Coteaux de la Loire**, **Muscadet Côtes de Grandlieu**, **Muscadet Sèvre et Maine**

Formerly, and still informally, known as Melon de Bourgogne, the Melon grape did originate in Burgundy, although its use there has been banned since at least the sixteenth century. Not only did the Burgundians reject the grape, but these days are quite litigious about others using the B word outside their region, so the grape is now officially simply Melon. After being banished from Burgundy, it arrived in the Nantais to a more favourable reception in the 1600s. Here it was valued by the Dutch as a source of light, acidic, neutral wines that were useful for distillation and subsequent transport overseas. At this time Folle Blanche vines outnumbered those of Melon, due to its higher acidity, but Melon became the preferred variety when its hardiness was proved during the devastating freeze of 1709 (reportedly Europe's coldest winter in 500 years).

The origin of the name 'Muscadet' is unclear, but it is possibly related to its mild musky aroma, or the fact that the Dutch used to blend it with spices such as nutmeg (*noix de muscade*). Nowadays it is the sole permitted variety for all Muscadet wines designated *sur lie* or *cru*. (Basic Muscadet wine is allowed to contain a small proportion of Chardonnay, but traditionalists generally reject such practice.) Production declined during the 1980s, but Muscadet is enjoying something of a revival, as ambitious producers have learned that long ageing on lees gives savoury wines of subtle, creamy complexity and undeniable longevity. The development of the *crus communaux* in the last twenty years has led to renewed interest in a grape variety capable of producing saline and terroir-specific wines which have natural acidity and low alcohol levels when grown in the oceanic climate of the Nantais. There are no significant plantings of Melon anywhere else in the world, so these wines are some of the unique treasures of the Loire.

Chardonnay – around 2,900 hectares

Cheverny, Coteaux du Vendômois, **Côtes d'Auvergne**, Crémant de Loire, Fiefs Vendéens, Muscadet, Orléans, Saint-Pourçain, Saumur Mousseux, Touraine-Mesland, Touraine sparkling, Valençay

The Loire's fourth most-planted white variety does not have a strong presence of its own compared to its popularity elsewhere in the world. Varietal Chardonnay wines, such as the small number found in the Côtes d'Auvergne, the Loire's only 100 per cent appellation for the grape, tend to be delicate and light in body. Chardonnay can be used as a minority blending partner for basic Muscadet, giving a little more interest and weight to an even lighter wine from Melon. Chardonnay is at its most impressive in the region as a sparkling wine, where it can add a lemony, creamy elegance to blends comprising less noble varieties.

Chardonnay is a minor grape in the Coteaux du Vendômois with Chenin, and in Cheverny with Sauvignon Blanc. It also features in blends in Orléans whites, and in Saint-Pourçain, where it complements the local speciality, Tressallier.

Other white grapes

Chasselas

Pouilly-sur-Loire

Chasselas is an ancient variety that was originally planted around Pouilly as a table grape to be sent to market in Paris. According to Sébastien Redde, of Domaine Michel Redde, 'Chasselas needs a lot of work to reduce yields and make good wine.' Theirs is one of the more interesting examples. Chasselas is a minority grape in Pouilly, where plantings are only in the tens of hectares. The wines are light-bodied with fairly neutral flavours and drunk when young and fresh.

Folle Blanche

Gros Plant du Pays Nantais

Also known as Gros Plant, this is a grape that originates in the Charentais region of France, where it was principally valued for its high acidity for *eau-de-vie* production. With the arrival in Nantes of the Dutch merchants for whom distillation was the priority for shipping back home, Folle Blanche was a natural choice. The first known mention of the grape being made into a still wine was in 1732, and it soon became a successful wine grape variety in its own right in western France. By the middle of the nineteenth century, even with the decline in distillation, Folle Blanche plantings in the Nantais stood at 20,000 hectares. At the time of writing, this has fallen to under 400 hectares in the Muscadet area, which is a decline of more than 1,000 hectares since 2009.

As the 'other' white wine of the Muscadet area, Folle Blanche produces wines that are, without exception, extremely acidic. They are slightly lighter in body than Muscadet, and often have an attractive floral scent and citrussy tang.

Menu Pineau/Orbois

Cheverny, Crémant de Loire, Touraine, Touraine sparkling, Valençay, Vouvray

Also known as Orbois, this is a local Loire variety with low acidity that is permitted as a minor part of blends. Formerly widely planted in Touraine, it has now largely been supplanted by Sauvignon Blanc. It is a minor blending partner in other appellations and can comprise up to 5 per cent of Vouvray wines.

Pinot Gris/Malvoisie

Châteaumeillant (rosé), **Coteaux d'Ancenis**, Orléans rosé, Reuilly, Touraine Noble-Joué, Touraine rosé/sparkling

Pinot Gris is little seen as a single varietal wine other than in the style of sweet Malvoisie in Coteaux d'Ancenis, which is the most-produced wine of the appellation. Most of this is consumed locally. Pinot Gris is also blended into interesting, fleshy rosés, known as *vins gris* because of their pale, onion-skin colour, in the appellations of Reuilly and Touraine-Noble-Joué.

Romorantin

Cour-Cheverny

Romorantin is a Loire speciality unique in the world to the appellation of Cour-Cheverny, where plantings of the grape number around 70 hectares. It is always a single varietal wine. Originally from Burgundy, but now firmly established as a Loire variety, Romorantin is a tricky grape to ripen, but can be rewardingly different in a good vintage and in the hands of a good winemaker.

Sacy/Tressallier

Saint-Pourçain

This rarely seen variety is locally known as Tressallier in Saint-Pourçain. It is vigorous and fertile, but when yields are kept in check it can produce a light, mineral, high-acid wine of real character, somewhat reminiscent of Chardonnay. To be entitled to the appellation it must be blended with Chardonnay, so some producers make varietal wines outside the Saint-Pourçain regulations. The grape fell from popularity due to its tendency to produce wine with high acidity and low alcohol, but as these characteristics are now more highly valued, it is possible that the mere 50 hectares planted with the variety (all in Saint-Pourçain) may increase in the coming years.

Sauvignon Gris

Cheverny, Haut-Poitou, Touraine, Valençay

The variety is used as a minor blending partner in these appellations.

The main black grapes

Cabernet Franc – around 16,000 hectares

Anjou, Anjou Brissac, Anjou-Villages, Bourgueil, Cabernet d'Anjou, Chinon, Coteaux d'Ancenis (up to 10 per cent), Coteaux du Loir, Coteaux du Vendômois, Crémant de Loire, Fiefs Vendéens, Haut-Poitou, **Orléans-Cléry,** Rosé d'Anjou,

Saint-Nicolas-de-Bourgueil, Saumur, Saumur-Champigny, Saumur Mousseux, Saumur Puy-Notre-Dame, Touraine, Touraine-Azay-le-Rideau rosé, Touraine-Chenonceaux, Touraine Gamay, Touraine-Mesland, Touraine rosé/sparkling, Valençay

Although red wines are in a minority in the Loire Valley, Cabernet Franc is the single most planted grape variety, featuring in a wide range of red, rosé and sparkling wines. It accounts for one in four vines (of any colour), and more than half of all red wine grapes. The Loire is the number one region in the world for Cabernet Franc. Although not native to the Loire (or, most likely, even to France, see p. 184), the winemakers of the Loire can surely claim Cabernet Franc as their own today, for it is here that it finds its finest expression as a varietal red wine. Cabernet Franc is the grape of choice partly from historical associations – it has been here for at least 500 years – but it is also a pragmatic choice in more recent times. The Loire is a marginal climate for ripening high-quality red wine grapes, and Cabernet Franc is an early ripening variety (compared to its partner and offspring, Cabernet Sauvignon, for example).

When ripe and carefully handled, Cabernet Franc gives deliciously aromatic wines with raspberry and violet fragrance and some of the familiar black fruit characters of Cabernet Sauvignon. If not fully ripe, or poorly managed, Cabernet Franc has a tendency to express leafy, vegetal characters. Historically, the Loire has suffered from a poor reputation for its Cabernet Francs for this reason, as well as the tendency to over-extract these green tannins. There are definitely some wines that continue this less than noble tradition. But increasingly, the modern face of Loire Cabernet Franc is quite different. With riper vintages and more modern winemaking, what used to be raw and green is now more likely to be ripe, fruit-forward and refreshing.

Despite its importance as the most-planted red wine variety of the Loire, only the minor appellation of Orléans-Cléry requires growers to use 100 per cent Cabernet Franc. In practice, however, it is usually used exclusively in the flagship appellations of Bourgueil, Chinon, Saint-Nicolas-de-Bourgueil and Saumur-Champigny, where the limestone-based soils allow the potential of the grape to shine through. The epitome of the fresh Cabernet Franc style is Saumur-Champigny, where the purity of its tuffeau soils allows the grapes to retain a cool freshness regardless of the vintage. These wines can have a quality of weightless elegance to them. Crisply appealing when young, the best can age well. The lighter examples are excellent served slightly chilled.

Opposite in style is Chinon, the largest red wine appellation, which produces succulent and deeply coloured Cabernet Francs. These often dense wines feature dark fruit characters such as black cherry and blueberry and the best need a few years to open. To the north of the Loire, the slightly cooler appellations of Bourgueil and Saint-Nicolas-de-Bourgueil produce wines in a somewhat lighter style. In particular, those of Saint-Nicolas are known for their easy-drinking, red-fruited character, while those of Bourgueil can be structured and elegant, especially when grown on the clay and limestone slopes. They feature flavours of cherries, raspberries and cassis.

A large proportion of the Loire's Cabernet Franc is used as a base for Cabernet d'Anjou, as well as the other high-volume rosé appellation, Rosé d'Anjou.

Gamay – around 4,200 hectares

Anjou (including **Anjou Gamay**), Châteaumeillant, Cheverny, Coteaux d'Ancenis (reds and rosés), Coteaux du Giennois, Coteaux du Loir, Coteaux du Vendômois, **Côte Roannaise**, Côtes d'Auvergne, **Côtes du Forez**, Fiefs Vendéens, Haut-Poitou, Rosé d'Anjou, Saint-Pourçain, Saumur Mousseux, Touraine, Touraine Amboise rosé, Touraine-Azay-le-Rideau rosé, Touraine Gamay, Touraine-Mesland, Touraine rosé/sparkling, Valençay

Gamay appears as a light, quaffable red wine in vineyards from the source to the estuary of the Loire. Almost one in five of France's Gamay vines finds its home in the Loire Valley. For the most part, Gamay wines are just a part of a vigneron's wider range, but the grape really comes into its own in the upper Loire. On the volcanic and granitic soils of the high-altitude Côte Roannaise and Côtes du Forez appellations, Gamay is the source of uniquely succulent and spicy wines. Here they grow a particular clone, Gamay Saint-Romain, whose vines produce smaller berries than the Gamay Beaujolais, giving deliciously juicy wines that have concentration despite their very moderate levels of alcohol. The same clone is found in blends in the nearby appellations of Côtes d'Auvergne and Saint-Pourçain.

At the opposite end of the Loire, the Coteaux d'Ancenis has attractive, youthful, raspberry-scented reds that are usually 100 per cent Gamay although they may include some Cabernet Franc. Touraine and Anjou have separate appellations for their crisply fruity Gamays, although the variety need only make up 85 per cent of the blend. Touraine-Mesland reds must contain a good proportion of Gamay too, blended alongside Cabernet Franc and Côt. The Touraine-Mesland rosés are mostly from Gamay as well.

Pinot Noir – 2,200 hectares

Châteaumeillant, Cheverny, Coteaux du Giennois, Coteaux du Vendômois, Côtes d'Auvergne, Crémant de Loire, Fiefs Vendéens, Haut-Poitou, **Menetou-Salon**, Orléans (reds and rosés), **Reuilly**, Saint-Pourçain, **Sancerre**, Saumur Mousseux, Touraine, Touraine Gamay, Touraine Noble-Joué, Touraine rosé/sparkling, Valençay

The Loire grows around 10 per cent of France's Pinot Noir. It is the most important quality red grape of the Centre Loire, but it has only a minor presence compared to the dominant Sauvignon Blanc. The best wines are extremely impressive, terroir-specific Pinots of silky depth and bright acidity.

Pinot Noir makes up the majority of Cheverny reds, where it is blended with Gamay and Côt. It is also blended with Gamay in Saint-Pourçain, where it is often known by its local name of Auvernat. In the Fiefs Vendéens it is also seen in blends with the seemingly incompatible Négrette variety, leading some producers to dispense with the appellation to make their Pinot Noirs under the Vin de France label.

Grolleau – around 2,000 hectares

Anjou, Coteaux du Loir, Crémant de Loire, Fiefs Vendéens, Rosé d'Anjou, Rosé de Loire, Saumur Mousseux, Touraine-Azay-le-Rideau rosé, Touraine rosé and sparkling

Though rarely produced as a varietal wine, Grolleau is the fourth most planted black grape of the Loire. It is a major component of Rosé d'Anjou and makes up the majority of the wines of Touraine-Azay-le-Rideau rosé. Along with its pink-skinned variant, Grolleau Gris, it is also a popular and productive choice for inexpensive sparkling wines.

Other black grapes

Cabernet Sauvignon

Anjou, Anjou Brissac, Anjou-Villages, Bourgueil (up to 10 per cent), Cabernet d'Anjou, Chinon (up to 10 per cent), Crémant de Loire, Fiefs Vendéens, Rosé d'Anjou, Saint-Nicolas-de-Bourgueil, Saumur, Saumur Puy-Notre-Dame, Touraine, Touraine-Azay-le-Rideau rosé, Touraine Gamay, Touraine rosé/sparkling

Far less planted than Cabernet Franc, Cabernet Sauvignon has only a complementary role to play in the wines of the Loire, where it is usually blended with Cabernet Franc. On the well-drained schist soils of Anjou it provides added structure and interest if it achieves full ripeness. It is also one of the two permitted grapes (the other being Cabernet Franc) in the Loire's largest appellation, Cabernet d'Anjou.

Côt/Malbec

Cheverny, Coteaux du Loir, Rosé d'Anjou, Touraine, **Touraine-Amboise red**, Touraine-Amboise rosé, Touraine-Azay-le-Rideau rosé, Touraine-Chenonceaux, Touraine Gamay, Touraine-Mesland, Touraine rosé/sparkling, Valençay

Côt, the local synonym for the Malbec grape, is important in Touraine, where it is the most important black grape, accounting for at least 40 per cent of red blends. It is more important in Touraine-Amboise, an unusually inky but highly approachable speciality that is 100 per cent Côt. It is also important for rosés in this appellation, and for Touraine-Chenonceaux reds where it makes up the majority of a blend with Cabernet Franc. Côt also appears in the red wines of Cheverny, where it is blended with Pinot Noir and Gamay.

Merlot

Haut-Poitou

Despite its popularity further south, Merlot is very rarely seen in the Loire.

Négrette

Fiefs Vendéens

Négrette is a traditional grape that must be included in the red wines of the Fiefs Vendéens. It is savoury in flavour, and blended here with Pinot Noir and Cabernet Franc.

Pineau d'Aunis

Anjou, Coteaux du Loir, Coteaux du Vendômois, Crémant de Loire, Rosé d'Anjou, Saumur, Saumur Mousseux, Saumur Puy-Notre Dame, Touraine, Touraine rosé/sparkling, Valençay rosé

This ancient variety is no relation to anything that sounds remotely similar. Highly productive, it would historically slip anonymously and generously into generic sparkling blends, but it is currently undergoing a minor revival in interest as a quality still red wine. The traditional region for this variety is the Loir Valley, north of the Loire. Here, in the Coteaux du Loir and Coteaux du Vendômois appellations it is the majority grape for red and rosé wines, and may also be made as a varietal wine. At low yields (and even better, when the vines are also old) these pale, surprisingly tannic wines are uniquely interesting, and have aromas of red fruits and white pepper.

Pinot Meunier

Orléans (reds and rosés), Touraine-Noble-Joué, Touraine rosé/sparkling

Pinot Meunier makes an appearance only in a couple of appellations. It must comprise the large majority of Orléans reds and rosés, along with Pinot Gris and Pinot Noir, and it is the principal component of the blend for Touraine-Noble-Joué wines, with Pinot Gris and Pinot Noir.

4

THE NANTAIS: OCEAN WINES

The winemakers of the Nantais live life on the edge. On the fringe of the European continent, the most westerly vineyard region in France is also at the northern limit of French wine growing; venture further north and you are in Brittany, apple and cider country. The oceanic climate may soften the effects of this northerly latitude, but spring frosts can be hard and unforgiving, and summers sunny but cool. Winemaking in the Nantais is an optimistic pursuit, an annual bet against the weather. Wine growers here are more resilient than most.

Dominated by the Muscadet appellations – more than 6,000 hectares of Melon, a grape that grows virtually nowhere else on the planet and the largest area devoted to a single white variety in France – the Nantais also includes 800 hectares of plantings across the AOCs of Gros Plant du Pays Nantais, Coteaux d'Ancenis and the Fiefs Vendéens. Still, dry white wines account for the vast majority of production in the Nantais region. Reds and rosés come almost entirely from the Coteaux d'Ancenis and Fiefs Vendéens appellations.

This is the Loire's last glimpse of vines before it empties into the Atlantic. Some of the vineyards reach as far as the coast, but the ocean's influence is felt throughout the area: it is in the air, the climate, and even the people. In an inescapably aqueous environment of rivers, lakes and marshes, the sky, too, provides its fair share of water, around 800 millimetres per year. Climate change has brought warmer growing seasons, which have aided the reliability of ripening – but at a cost. Precocious early-season growth coupled with late frosts in spring can be devastating, especially in a region where average per-bottle prices remain low.

All this means it has been difficult to attract new entrants into the business, or to retain the interest of the next generation. The attractive rolling green countryside of the Muscadet appellation, most of which is within striking distance of the city of Nantes, often has greater value for housing development than for vineyards. The same can be said for the land close to the coast, where changes in office working practices post-Covid have led to rocketing prices. Many of the coastal vineyards of the Fiefs Vendéens have been abandoned, but owners are reluctant to sell while prices are still on the rise.

In some appellations, growers have been active in recent years in lobbying for changes to allow greater flexibility to deal with difficult climatic conditions. In the Fiefs Vendéens, rules have been relaxed regarding the precise percentages of the varieties that must be included in these blended wines, which helps them to deal with vagaries in weather and ripening. In Muscadet, the most recent quality development of the Crus Communaux has provided growers with added insurance of sorts, since in lean years they at least have stocks of their best wines ageing slowly underground. This is a practice that has been commonplace for years outside the confines of the appellation, but that it is now officially recognized adds value to these unique and age-worthy wines.

Nantais – key facts

- **Hectares in production**: 7,600 (13 per cent of Loire total)
- **Annual production volume**: 300,000 hectolitres/40 million bottles (11 per cent of total Loire production)
- **Styles**: 95 per cent white, 2 per cent rosé, 3 per cent red

AOCs of the Nantais

- Muscadet
- Muscadet Sèvre et Maine
- Muscadet Coteaux de la Loire
- Muscadet Côtes de Grandlieu
- Gros Plant du Pays Nantais
- Coteaux d'Ancenis
- Fiefs Vendéens

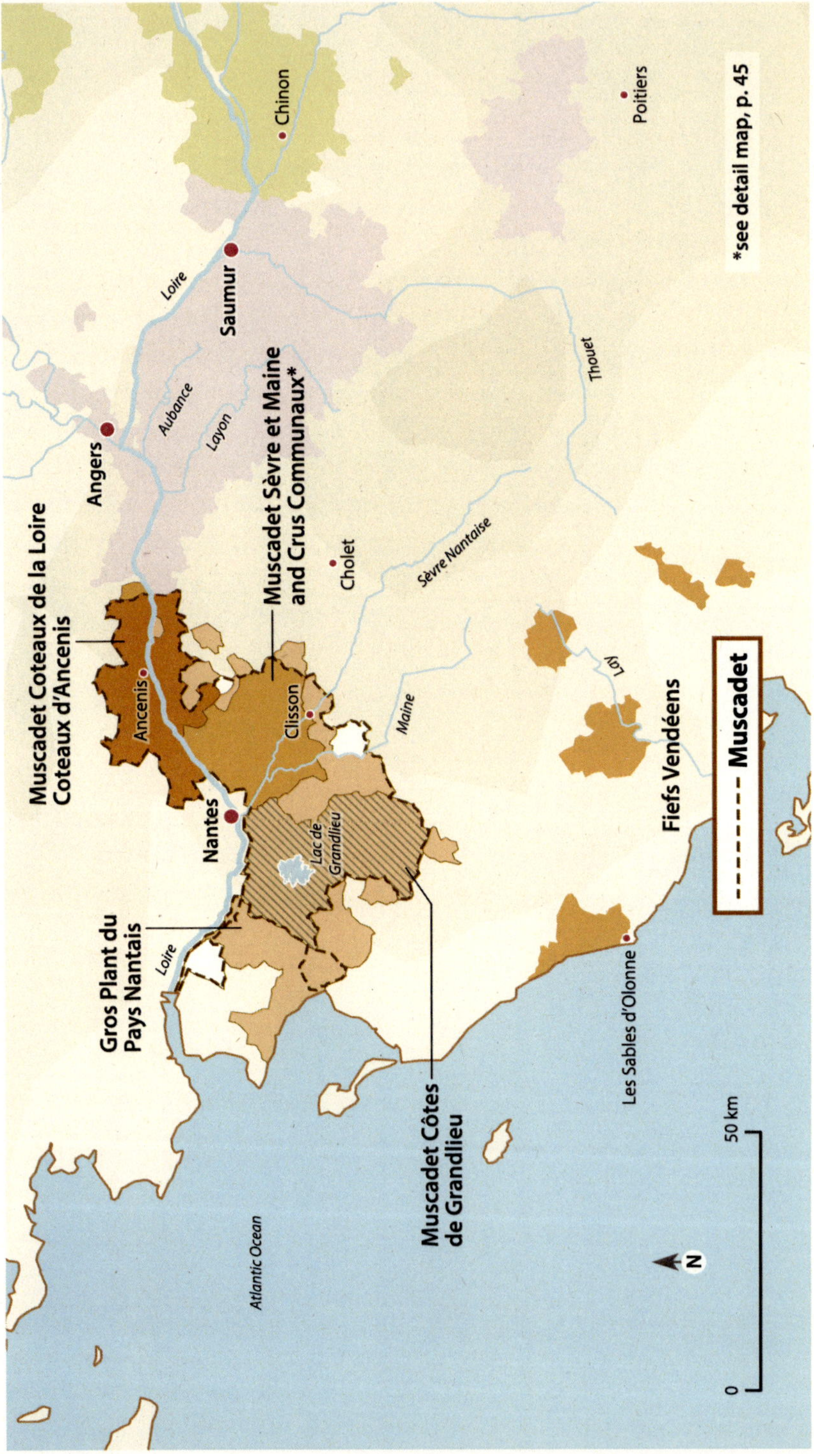

Map 3: Nantais

The Muscadet appellations – key facts

- **AOCs granted**: 1936, 1937, 1994
- **Hectares in production**: 6,713
  Muscadet: 1,272
  Muscadet Sèvre et Maine: 5,010
  Muscadet Sèvre et Maine Crus Communaux: 103
  Muscadet Coteaux de la Loire: 106
  Muscadet Côtes de Grandlieu: 222
- **Annual production volume**: 262,000 hectolitres/35 million bottles
- **Grapes**: Melon, Chardonnay (the latter permitted only for basic AOC Muscadet)
- **Styles**: dry white
- **Soils**: varied soils of the Armorican Massif: gneiss, gabbro, granite, amphibolite, micaschists

MUSCADET: MELON AS FAR AS THE EYE CAN SEE

One of France's earliest designated appellations, Muscadet covers a broad area to the east, south and south-west of the city of Nantes. In many parts of the Loire Valley, one has to look quite hard to find the vineyards, but not here: vines dominate the landscape. Despite the obvious concentration of vines, though, the Loire's single largest white wine appellation has been steadily declining in plantings and output for decades. In the early 1990s annual production of Muscadet was two and a half times what it is today, although even at this point Muscadet's fall from grace was already well underway. In the 1980s the region enjoyed a boom in exports, especially to the UK, where Muscadet was valued as a reliable and inexpensive light white wine to slake the considerable thirst of the British market. But as plantings increased to meet demand, yields increased also, with an inevitable decline in quality. At the same time, prices rose as merchants raced to keep up with demand. This unsustainable situation came to a head with the drastic and devastating frosts of 1991 that destroyed most of the Muscadet crop. Their customers were forced to look elsewhere. Unfortunately for the growers, this period coincided with the rise in exports to the UK from Australia,

New Zealand and other warmer-climate regions which provided a more reliable source of good-value wines bursting with fruit and sunshine. Muscadet simply couldn't compete.

Today, however, the picture looks somewhat brighter and, at least from a buyer's perspective, there are plenty of reasons to be cheerful about Muscadet. Wines that might have been considered old-fashioned in the 1990s are looking increasingly modern in the twenty-first century. In a world where winemakers are often struggling to control ripeness and alcohol levels, Muscadet excels: 12% abv is the norm for these wines, and even in the warmest of recent vintages, the wines stay under 13% abv. Nor is acidity a problem in this cool, maritime climate: Muscadet is always refreshing. Naturally light, crisp wines used to be easy to find, but are today increasingly hard to come by. While winemakers in warmer climates search for new, cooler, or higher altitude places to plant, this historic region continues to be a source of naturally light, dry, savoury wines. What Muscadet lacks in brashness, it makes up for in subtlety and sheer drinkability. And the best of these wines can age for decades. When it comes to food, Muscadet – in particular, Cru Muscadet – has the edge over most white wines due to its extended ageing on lees. This gives these wines a savoury, umami character that makes them wonderful partners for many kinds of food. Even if the quantity of Muscadet in the world is declining, there is an increased focus on the quality of the wines available to enjoy.

Muscadet comprises several appellations, but essentially it promises and delivers a single style from its unique grape variety: still, light, dry white wine. And at a fair price. On reflection, this unusually simple branding may have contributed much to its overseas success. If you buy a bottle of Muscadet, you know exactly the style of wine you are going to get. Incidentally, 'Muscadet' is a synonym for the Melon grape; there is no place called Muscadet.

The area of the basic appellation's production encompasses all the Muscadet appellations and accounts for just over 1,000 hectares of the region's output. There are two distinguishing differences between basic Muscadet and the other appellations. The first is the permitted volume. Muscadet can be made with yields of up to 70 hectolitres per hectare, compared to 55 hectolitres for the other Muscadet appellations. At production levels as high as this Melon – not the shoutiest of personalities at the best of times – produces pretty neutral, if inoffensive wines. The second difference is that for the basic appellation, up to 10 per cent of

Chardonnay can be blended into the Melon for extra weight, texture, and perhaps a modicum of added interest. This addition is not routinely made, however, and this is the only one of the Muscadet appellations that allows the inclusion of a second variety. Many locals frown upon this intrusion into the monovarietal landscape of Muscadet, and any other wine with Muscadet on its label is guaranteed to be 100 per cent Melon.

The most important of the appellations is Muscadet Sèvre et Maine, which at 5,000 hectares amounts to three-quarters of the vineyard area. The vines grow in a fairly compact area south of the Loire, around the tributaries of the Sèvre Nantaise and the Maine. Here yields are restricted to 55 hectolitres per hectare and are commonly labelled '*sur lie*' (a designation not permitted for basic Muscadet).

The practice of ageing wine on lees dates back to the beginning of the twentieth century, when wine growers often put aside their best barrel to age over the winter for future family celebrations. They discovered that the yeast lees remaining after fermentation conserved the wine's freshness, adding a layer of complexity, a lightly mealy texture and often a spritz of carbon dioxide. In 1977 the INAO formalized this practice by allowing growers to add the words '*sur lie*' to Muscadet labels if the wines had been aged in this way and bottling had been delayed until 1 March following the harvest. So far, so logical. But this is where the logic ends, because there is a limit to the amount of time the wine can be left on the lees to be entitled to the *sur lie* qualification. This period is between five and 14 months. This means that wines left for longer than the end of the year following the fermentation are not permitted to use the *sur lie* declaration, even though these wines will show considerably more lees character the longer they are left in contact with the lees (think about vintage Champagne versus non-vintage: much of the character and quality of vintage Champagne can be attributed to the longer lees ageing required for this style of wine). In reality, the Muscadets showing the most lees character are often not the ones labelled *sur lie.* Instead, these wines often fall into the newest, and most interesting category of Muscadet wines: the Crus Communaux. However, in recognition of the highly confusing labelling law regarding lees, from 2025 '*sur lie*' will be replaced by the words '*élevé sur lie*', meaning 'matured on lees', followed by the amount of time in months that the wine has been kept on lees. Which should, hopefully, be less confusing all round.

The ten Muscadet *crus* represent a mere two per cent of all Muscadet production, but are leading the way in terms of quality. They have highlighted the potential of the humble Melon grape for expressing the diverse terroirs of the region. Beneath the soils of Muscadet lie the highly complex, igneous and metamorphic rocks of the Armorican Massif. These ancient soils date back to the primary era, some 450 million years ago, and include a complex mix of gabbro, granite, gneiss and amphibolite, among others.

Over the years, growers in the Sèvre et Maine Muscadet heartland had noticed differences between wines grown on these various soil types and some began to separate their vinifications according to the different soils. They also began to experiment with longer periods of lees ageing – years rather than months. Convinced by the results, they applied for special status from the INAO. After the usual protracted procedures, their efforts were rewarded around a decade later with the official recognition of Crus Communaux. It should be noted that these *crus* are not separate appellations in themselves, but what is known as Dénominations Géographiques Complémentaires (or DGCs). These are geographical indications that can be used on the labels in addition to the appellation, for example, Muscadet Sèvre et Maine Clisson ('*cru*' or 'DGC' are not terms one will see on any label). The *cru* wines have a lower maximum yield limit of 45 hectolitres per hectare, instead of 55 for the sub-regional Muscadet appellations. They cannot be made with the fruit of very young vines, although the youngest acceptable vine age is still quite low, at six years. Each *cru* has its prescribed minimum period for ageing on lees, which is typically between 18 months and two years, but can be as long as three years.

The first *crus* to be recognized were Clisson, Gorges and Le Pallet, in 2011, followed in 2019 by Goulaine, Château-Thébaud, Monnières-Saint-Fiacre and Mouzillon-Tillières. At the time of writing, the final three, La Haye-Fouassière, Vallet and Champtoceaux were awaiting approval, but there is no reason to suppose they will be denied *cru* status. The *crus* are cultivated over an area of around 200 hectares, all within the Sèvre et Maine appellation except for Champtoceaux, which is in the Coteaux de la Loire region to the east of Nantes.

Cru communal wines are made by around 100 growers. It would take a real expert to distinguish reliably between the wines from all the different *crus* if tasted blind, but there are distinct differences discernible between many. As with all appellations, often the hand of the grower is

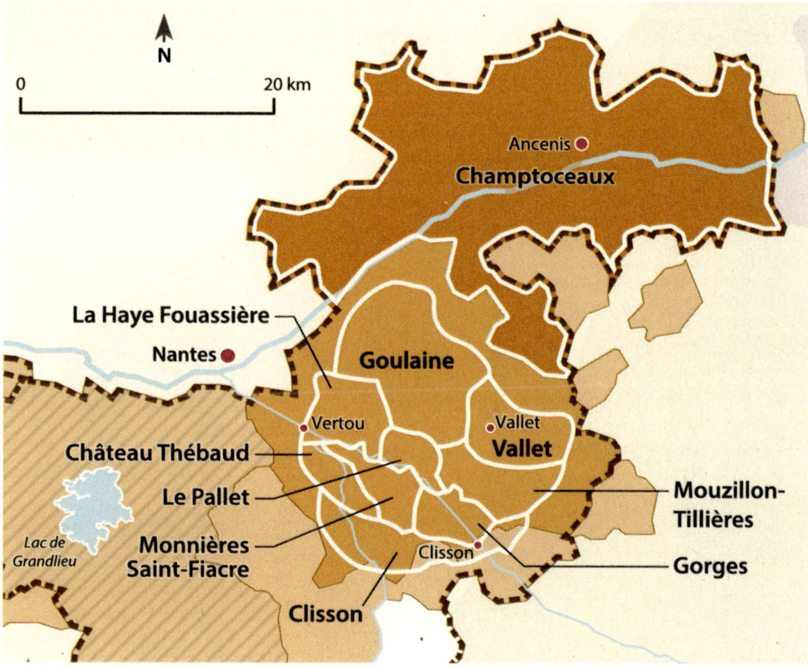

Map 4: The Muscadet Crus Communaux

as evident as the *cru* – for example, the extent of lees movement (*bâtonnage*) employed by the vigneron will significantly affect the style of the wine. Furthermore, because the winemakers are often still experimenting with these relatively new wine styles, some producers will age their wines for longer than others before release. There is no maximum ageing period, and some wines are aged for up to a decade. Lees ageing is part of what makes these wines so interesting and adds to their unique appeal. There is no doubt that they represent the pinnacle of Muscadet production today. Below is a summary of the main characteristics of the *crus*.

Getting to know the Crus Communaux

Clisson: Free-draining and particularly heat-retaining granite soils give the ripest and most powerful wines of all the *crus*, often with up to a degree more alcohol than other Muscadet wines. Open, floral, aromatically fine and persistent, these are perhaps the easiest of the *cru* wines to appreciate, especially when young. As a result, this is probably the most demanded and most well-known of the *crus*. Centred on the lovely Italian-styled town of Clisson, the vineyards lie on a fault line of granite. These are gravelly, well-drained soils that provide ideal ripening conditions. The wines must be aged on lees for a minimum of 24 months.

Gorges: Situated on both sides of the Sèvre Nantaise river, Gorges vines grow on altered gabbros, clay and quartz soils that retain water and remain cool compared to the warmer granites of Clisson. The wines age for two years on lees, and have flavours of menthol, citrus, stone and smoke. Gorges wines are firm and savoury, with intense minerality and sometimes a resinous character. Razor-sharp, they can be unyielding in their youth, but are capable of ageing for decades. Probably the longest-lived *cru*, and opposite in style to neighbouring Clisson, these are serious, impressive wines that need time to reveal themselves.

Le Pallet: On the right bank of the Sèvre, the Le Pallet *cru* has warm conditions that give earlier-ripening grapes. Soils are shallow and stony. Wines must spend 18 months minimum on lees, giving them a smooth character.

Goulaine: Situated around the Goulaine wetland area, the vines ripen easily in deep, sandy soils over siliceous metamorphic rock, including gneiss and mica schist. They must be aged for 20–30 months on lees. Wines often have a silky texture and flavours of honey and ripe fruit, white pepper and savoury notes.

Château-Thébaud: Multiple soil types abound in the Château-Thébaud *cru*, but all are shallow. They include sandy soils over subsoils of granodiorite and gneiss. The best wines of this *cru* combine the more extreme characteristics of other *crus* to give firm and complete, smoky, mineral, creamy, roundly textured, and extremely attractive wines. They may feature flavours of fennel and liquorice and are matured for 36 to 48 months on lees.

Monnières-Saint Fiacre: Close to the village of the same name, on the left bank of the Sèvre, vines grow on sloping vineyards in moderately deep, fine, silty sand or decomposed gneiss with clay. These wines can be full-bodied (in the context of Muscadet) and creamy, with a mineral bite. The best examples offer delicacy, finesse and a lacy texture.

Mouzillon-Tillières: On the banks of the Sanguèze in eastern Sèvre et Maine, the vines are situated on moderately deep soils, with subsoils of coarse-grained gabbros. The wines feature citrus flavours and white orchard fruits (apples, pears), and become broader and more open with age. The wines age for two years or more on lees.

La Haye-Fouassière: This is the westernmost of the *crus*, featuring shallow, pebbly soils and a subsoil of orthogneiss. Grapes mature quickly here, giving expressive wines, with menthol, mineral, flint and floral notes. After 18 months minimum on lees, these wines may mature more quickly than other *crus*.

Vallet: Situated in the eastern part of the region, with shallow to moderate sandy/clay soils and a subsoil of mica schist, the wines of Vallet are relatively rich, elegant and expressive. They age for 18 months or more on lees.

Champtoceaux: The youngest of the *crus* and the only one to be situated outside the Sèvre et Maine appellation, Champtoceaux is in the appellation of Muscadet Coteaux de la Loire, on the hills around Ancenis. This is open land with optimal sun exposure. Strong winds from the Loire create a healthy growing environment for the grapes, favouring high levels of ripeness. Wines are silky and textured, with more noticeable weight and ripeness than the other *crus*. Scents of orange blossom and fresh fruits (citrus, peach) dominate in these wines, which are aged for a minimum of 17 months on lees.

Muscadet Coteaux de la Loire and Muscadet Côtes de Grandlieu

The two smaller Muscadet appellations lie to the east and west of the Sèvre et Maine. The boundaries of Muscadet Coteaux de la Loire mirror those of the Coteaux d'Ancenis appellation, with only around 100 hectares devoted to Muscadet production. Vines grow on open hillsides on both sides of the Loire, around the pretty town of Ancenis. Proximity to the river gives the grapes plenty of aeration and the exposed hillsides accelerate ripening. Muscadets from the Coteaux de la Loire tend to be riper and fruitier in style compared with those from further west. In contrast, the 222-hectare Muscadet Côtes de Grandlieu to the west is more influenced by the ocean. A mere 20 minutes from Nantes airport (and easily visible from the air when flying in or out of the city), vines grow within the urban conurbation of Nantes, to the north and south of the 6000-hectare nature reserve of Lac de Grand-Lieu. The lake covers a large area, but the space dedicated to grapes is small. The low-lying vines to the north of the lake are frost-prone, while the southerly vineyards have less frost but more rain. This is a challenging climate for viticulture, even within the Muscadet region.

GROS PLANT DU PAYS NANTAIS

This appellation covers an area encompassing the entire Nantais region. Gros Plant, usually known by the more attractive name of Folle Blanche, used to be the dominant grape variety back in the 1700s, when its abilities as a productive, high-acid grape were prized by the Dutch to produce a neutral base for brandy to be shipped to Holland. Even though Folle Blanche plantings have now diminished and are tiny compared to those of Melon, the area nonetheless produces nearly three million bottles of this fragrant, brisk, light white every year. Permitted yields are as for generic Muscadet, at 70 hectolitres per hectare. The acidity of Folle Blanche can be eye-watering, but in the right hands (in other words, those of a skilled Muscadet producer) and in the right vintage, it can be an uplifting and attractively floral wine. An especially good match for the briny local oysters, it also benefits from having even less alcohol than the average Muscadet.

Gros Plant du Pays Nantais – key facts

- **AOC granted**: 2011
- **Hectares in production**: 383
- **Annual production volume**: 19,000 hectolitres/2.5 million bottles
- **Grapes**: Folle Blanche
- **Styles**: dry white, some aged *sur lie* until the spring following the harvest.
- **Soils**: gneiss, mica schist, greenstone (gabbros and amphibolite)

Notable Muscadet producers

Domaine Brégeon

Gorges

www.domainebregeon.com

On the side of a fermentation tank at Domaine Brégeon is a quotation from the late oenologist and philosopher Jacques Puisais: '*Un vin doit avoir la tête de l'endroit où il est né et les tripes de celui qui l'a fait.*' It means, 'A wine should have the face of the place where it was born and the guts of the person who made it.' It seems an appropriate metaphor for the wines of Fred Lailler, and indeed the man himself,

The impenetrable gabbro rock of Gorges

a Gorges native who sports a bushy beard as black as the gabbro rock of his home.

Fred took over the domaine from Michel Brégeon in 2011. He farms 9 hectares of Melon and a single hectare of Folle Blanche. A small proportion of the vines grow on the granite of Clisson, but the rest are on gabbro, the defining face of the Gorges *cru*. Fred explains the qualities of the rock: 'It is denser, harder and less friable than Clisson granite. The topsoil has more clay, which makes it a cold soil. Gabbro soils don't suffer from water stress so the wines have more acidity.' These cool sites create some of the longest-lived Muscadets, but the wines can be austere, firm and uncompromising in their youth. 'Gorges wines should be like gunflint,' says Fred, clenching his fist. His approach to winemaking is as strict as his wines. Viticulture is organic, yields are low and everything is harvested by hand. The berries are pressed within two hours of picking and intervention is minimal. He says he aims to make wines that are 'fresh, light, and mineral,' and in this he succeeds, admirably, although he winces visibly when tasting a wine from the ripe 2018 vintage at 12.5% abv ('It's too much!' – the usual here is 12%). The *cru* wines can spend more than four years ageing on lees, but there is not a trace of fleshiness or roundness to Fred Lailler's wines because, unlike many producers, he never stirs the lees, dismissing the practice of bâtonnage as 'useless and violent.' The result is wines that have focus and intensity allied to a saline freshness and incisive acidity: wines that have the face of Gorges, and the heart and soul of Fred Lailler.

Domaine de la Pépière

Maisdon sur Sèvre

www.domainedelapepiere.com

This domaine was set up by Marc Ollivier in 1984 when he took over some family vineyards and rented a few more. The business grew and the wines soon found success abroad. When it was time for him to retire, in 2011, having no family of his own he handed the management of his business to his partner, Rémi Granger. Two years later Rémi was joined by Gwénaëlle Croix, who left a career in the agri-food business to pursue the life of a vigneronne. They now farm 43 hectares. Everything is biodynamic and harvested by hand.

Rémi Granger opens his cru wines at Domaine de la Pépière

The unassuming entrance of Domaine de la Pépière

The wines age in underground glass-lined vats in multiple locations which they rent from locals. There are a lot of spare vats hanging around in Muscadet, a throwback to when the vineyard area was so much larger than it is today. On a visit to the domaine, we drive from place to place for Rémi to lift a lid in what appears to be someone's garage floor and draw various pale, young wines into the daylight to taste. A young Clisson shows bright, zesty fruit. Château-Thébaud, also from a tank, is floral, with high acidity and a textured finish. Another Clisson shows more breadth in the mouth than the first. A vintage characteristic, I wonder? 'No,' say Rémi, 'It's the *bâtonnage*.' It's a delicate balance, he explains. 'During the ageing process [typically two years here], we're looking for the loss of primary aromas and we want an impression of weight, but not too much.' To achieve this he stirs the lees usually only once a year. They also store wine at the tiny domaine in fibreglass and stainless steel tanks, and concrete eggs.

The wines are understated, balanced, pristine and elegant. Their *cru* wines from Clisson, Monnières-Saint Fiacre, Gorges and Château-Thébaud are no less than shining representations of the terroirs of the *crus* of Muscadet. There's nothing showy here, and no excess of anything. Gwénaëlle and Rémi don't say much, but they don't really need to. Their wines speak for themselves.

Domaine du Haut Bourg

Bouaye
www.hautbourg.fr

Muscadet Côtes de Grandlieu was created in 1994. It is the youngest and most westerly of the Muscadet appellations. Formerly a part of the generic Muscadet classification, it covers a large area but plantings are relatively few. Domaine du Haut Bourg is found in the small Nantes suburb of Bouaye, close to the airport. Here, brothers Nicolas and Hervé Choblet manage the domaine their great-grandfather created in 1944. The name derives from its position at the high point of the town which, Nicolas informs me with a smile, is 15 metres above sea level. The brothers are a jovial pair and it strikes me that an ability to retain a light-hearted approach to life is probably a necessity when making wine in an always-uncertain and often-unfavourable climate.

With the Loire to the north and the large nature reserve of Lac de Grand-Lieu to the south, the air warms more quickly here than in the more densely planted Muscadet regions to the east, and seasonal growth starts earlier. But this makes the vines more vulnerable to spring frosts, which are even more difficult to deal with here than elsewhere, as there is understandable public resistance to the installation of wind turbines in the vines due to the inevitable noise pollution in a semi-urban environment. The humidity of the air from the watery environment (the rain, the Loire, the lake, the Atlantic …) also makes for high mildew pressure. Nonetheless, Nicolas and Hervé have managed to convert their best plots to organic viticulture.

The domaine has 38 hectares of vines, of which around half constitutes Muscadet. The brothers also produce varietal Gros Plant and a range of wines under the IGP Val de Loire appellation, including Sauvignon Blanc and still and sparkling Grolleau Gris. These are joyous, gulpable, everyday wines, many of which are snapped up by local passing trade, wine shops and restaurants. The Muscadets show purity and interest, especially those coming from the domaine's oldest vines. Of particular note is the cuvée Origine du Haut Bourg, which comes from the original hectare of vines planted by the brothers' great-grandfather in 1944. Aged for nine years in subterranean vats prior to release, it is balanced and yeasty, savoury and saline – and ages beautifully.

At the time of writing, plans were afoot among some growers, including Nicolas and Hervé, to apply for Côtes de Grandlieu to be added to the list of Muscadet *crus*, although many here sell direct to the public and don't see the interest. Whatever the outcome, one thing is certain: it will be a long process.

Domaines Joseph Landron

La Haye-Fouassière

www.domaines-landron.fr

Climbing into his dust-filled van, Jo Landron proudly announces that it has 360,000 kilometres on the clock – and that it is a new purchase. Landron, a slight figure with firm views and a twinkle in his eye, sports a moustache that can only be described as a statement. He also makes very fine Muscadet and has been practising his craft in La Haye-Fouassière since 1979. He farms 48 hectares around the winery, which is situated in a dominant position on the aptly-named Route de Bellevue. This high point – a mere 70 metres above sea level – is visible for miles around thanks to the gaily painted water tower that stands next to the winery. From the middle of his vineyard Jo points out the rock formations that underpin the Muscadet vineyards below: orthogneiss and granite to the north, gabbro and schist to the east near Vallet; further south towards Clisson and Gorges, degraded granite and mixed gabbro and granite. Under our feet is a narrow 150-metre vein of amphibolite, he informs me, the rest of the subsoil being orthogneiss.

Spend any time in the Muscadet vineyards and you realize the wine growers here have something of an obsession with rocks. Jo Landron is no exception, but he was something of a pioneer in identifying the effect of the different soils on his grapes and bottling site-specific, *lieu-dit* wines according to soil type. 'I realized this back in the 1980s,' he says. 'For example, amphibolite gives a straight, salty wine. Granite, on the other hand, gives wines that are aromatic, warmer and flinty.'

Jo has been farming organically since the turn of the century and the domaine has been certified biodynamic since 2008. Much of his approach to viticulture results from his many years of experience, intuition and paying attention to the vines, or as he succinctly puts it, *le bon sens vigneron*. 'It's a question of understanding the rhythm of the vine and respecting the plant,' he says.

Jo Landron's wines are models of purity and precision, born of a deep understanding of vine and place. Over the years, he says, he has learned to do less, to allow the wines to speak for themselves rather than to embellish them. Amphibolite is a striking example of this: a mouthwatering wine of tension and salinity with only around six months on lees. Taut yet easy to drink, it fairly floats across the palate, leaving a salty tang on the finish. Clos la Carizière comes from 50- to 70-year-old vines grown on orthogneiss. Perfumed and gentle, with layers of pear and apple fruit, it has eight months on lees. Les Houx is a mixture of grapes from two parcels on clayey, stony soil, an upright wine with sweet citrus notes and a touch of bitter grapefruit. His flagship wine, Le Fief du Breil, spends up to three years on lees. The south-facing, sloping plot from which the wine takes its name has old vines that grow on quartz and orthogneiss. This is a ripe, flinty and smoky wine with great concentration and mineral depth of flavour that remains nevertheless elegant and light. It ages almost imperceptibly: a ten-year-old can be as lively as a three-year-old and the wine proceeds magnificently into older age. A twenty-year-old example showed delicate aromatics of quince, dried apricots and honey and on the palate an explosive mineral intensity.

At the time of writing, the La Haye-Fouassière *cru* was awaiting final approval, but one gets the feeling that for Jo Landron it's all rather academic: 'The *lieux-dits* were recognized long before anyone thought of having a *cru* system,' he says, 'I've been making wines like this for 30 years already.'

Domaine Le Fay d'Homme

Monnières

www.lefaydhomme.fr

The interior of Vincent Caillé's winery gives an impression of barely controlled chaos, but nothing could be further from the truth. For it is here that this reflective and softly-spoken winemaker works with the utmost attention to detail to produce some of the best wines in the area.

Vincent was destined to be a winemaker. The fifth generation of a long line of Muscadet producers on both his mother's and father's sides, he took over the family's vineyards in 1986. From the outset, his focus was on quality. The estate was converted to organic viticulture in 2007 and certified biodynamic in 2012. He produces more than a dozen separate cuvées from his 25 hectares of vines, including a sparkling *méthode ancestrale* and four Vin de France reds from Gamay, Cabernet Franc, Côt and Abouriou. But it is his Muscadets that are the most notable wines of the collection.

His vines are spread among four communes and three main soil types. Most of the land consists of slopes of gneiss and orthogneiss, words which feature on some of his labels. He also has vines on the clay-rich gabbro terroir in the Cru Communal of Gorges. Everything is hand-harvested and yields are a low 35 hectolitres per hectare. Fermentations take place using only indigenous yeasts in traditional glass-lined cement tanks below ground. 'They are neutral and give temperature stability,' explains Vincent, 'and the cement between the tiles allows the wine to breathe.' After fermentation the wines are left to age on their lees, mostly undisturbed. '*Bâtonnage* brings oxygen,' he says, 'and Melon oxidizes easily, so you need to sulphite the wine every time you do this.' This is a point well illustrated by the two wines he makes that age in a concrete egg, Terre de Gneiss and Terre de Gabbro. The form of the egg means that the lees are in almost constant motion. This means their effects are felt more quickly, giving a more immediate and richer style of wine than others in the range.

The vinous highlights of the domaine are the *cru* wines from Monnières-Saint Fiacre and Gorges. These age on lees for up to 45 months and are supremely elegant expressions of the Muscadet terroir. Vincent's view of the two *crus* is that Monnières-Saint Fiacre gives finesse and precision, while Gorges has more acid tension. Both are delicious, but even the humbler wines of Muscadet Sèvre et Maine from this domaine are enormously expressive. The Fief des Coteaux Gabbro is a wine aged on lees, but Vincent chooses not to claim '*sur lie*' on the label in case he decides to age it for longer than a year, in which case it would fall outside the permitted ageing period for this classification. This fragrantly floral wine, from 60-year-old vines, has enormous energy, the linear acidity balanced by lime zest, confit lemon and ripe, crisp apple fruit. It is far from a simple Muscadet, with a long, mouthwatering finish. Two other Muscadet Sèvre et Maine wines reflect the other two types of terroir. Clos de la Févrie Orthogneiss comes from grapes grown on thin soils that warm quickly and drain easily. This is a riper style with baked apple fruit that manages to be both mouth-filling and light, with a light spiciness on the finish. Le Seigneur Gneiss is a quiet wine in its youth, with a salty, savoury character. It needs time in bottle to develop.

Vincent Caillé produces an array of beautiful wines that combine purity, delicacy and finesse, year after year. Constantly questioning how he works, he says he has only ever been satisfied with one vintage: 1996. But perhaps the best is yet to come.

Domaine Louis Métaireau Grand-Mouton

Saint-Fiacre

www.muscadet-grandmouton.com

Second-career vignerons Laure and Julien Rossignol, both agricultural engineers by training, bought this well-established domaine in 2019. Julien had spent the previous twenty years far from the land, in a commercial environment in Paris, where he worked in IT for Galeries Lafayette. When he decided it was time for a change of life for the family, Muscadet appealed for its affordability and its location close to the thriving city of Nantes. And, says Julien, they wanted somewhere where they could make their mark. He struck an agreement with the previous owners, Marie-Luce Métaireau and Jean-François Guilbaud, for a gradual handover, took a course in viticulture and winemaking and set to farming his 23-hectare Grand Mouton vineyard in Saint-Fiacre-sur-Maine. Dealing with the challenge of a pandemic was probably the last thing on his mind, but at least his commercial background helped in getting his online sales off the ground.

It is easy to see how one might be tempted out of the city to this lovely site: a house and winery set amidst the vines on a gentle mound (hence the 'mouton' – there are no sheep here) that slopes perfectly down to the nearby river Maine on one side and across towards the Sèvre Nantaise river on the other. Most of the domaine's vines are old, planted between 1937 and 1942. This was a period when vines were selected more for productivity than quality, but at their current ripe old age, they have the benefit of not only giving high-quality old-vine fruit, but also providing greater resistance to the vine trunk disease esca, which is more prevalent among younger vines.

Julien is converting the domaine to organic viticulture, a difficult prospect in Muscadet. He explains that in a year like 2021, due to heavy rain during much of the growing season it was necessary to make up to three tractor trips a week into the vines to spray copper sulphate against mildew. 'At some point,' he says, 'I wondered if we weren't going backwards with this approach.' In a year of extreme weather, it is hard to argue with his logic.

The wines of the domaine were greatly improved by the pioneering Louis Métaireau in the 1970s when he slashed yields by half in a quest for quality. Since this time, average quality in the region has improved as others have followed suit, so today the wines here are perhaps less remarkable for their quality than they once were. Various cuvées are produced according to the date of the harvest, which is all done by hand. Cuvée One is the flagship wine, made from the oldest vines from grapes harvested on the first day of picking and aged for around eight months on lees. Grand Mouton is picked later, and Carte Noir comes from grapes picked on the last day of harvest. In 2018 the domaine saw the introduction of the first wine to be aged for an extended time on lees. Named Collection 37, it is the same wine as Cuvée One, but aged for 24 months on lees.

In 2023 the domaine launched its first *cru* wine, Monnières-Saint Fiacre, a wine that comes from the oldest vines of the domaine and ages for 30 months on lees. Julien is also experimenting with fermenting in amphorae, and has created two wines using these vessels, including an orange wine. He is succeeding in making his mark on this historic domaine.

Domaine Luneau-Papin

Le Landreau

www.domaineluneaupapin.com

Pierre-Marie Luneau is the ninth generation to work his family's land in Muscadet. The Luneau-Papin domaine was formed in the 1970s when two winemaking families joined together with the marriage of Pierre Luneau to Monique Papin. Today it is Pierre-Marie and his wife Marie Cartier-Luneau who run this 35-hectare domaine. Nominally, Pierre is in charge of the wine and Marie the business side, but in reality, it's all hands to the pump on a daily basis. 'You have to be strong-minded to do this work,' says Marie Luneau, when I meet her during harvest (all done by hand). 'Strong and crazy,' she adds.

Pierre-Marie took over from his parents in 2005 and was joined by Marie in 2008. They have continued and improved on the work set in motion by the previous generation, separating their land into parcels according to soil type. As there are such varied soil types here, they make a lot of wines: twelve in all. Apart from a single Gros Plant wine, there are nine Muscadets and two Muscadet *cru* Goulaine wines. Harvest is manual and wines age in a mixture of amphorae and

Pierre-Marie Luneau, with a selection of his aged Excelsior wines

sandstone jars, wood (barriques, vats) and underground wooden vats. The quality is consistently good here. Terre de Pierre comes from vines planted in 1974. It is flinty, aromatic and salty, with fruity acidity, and ages effortlessly. The *cru* wines, Cuvée Excelsior and Gula Ana, are wines of pure pleasure. From vines that grow on the Butte de la Roche above the Goulaine wetlands, this is the earliest-ripening of the Muscadet *crus*. The wines are smooth-textured, silky and generous. Gula Ana was introduced in 2016. It is from vines that grow on the top of the hill on a subsoil of serpentinite. It is powerful, mineral and satisfying. Cuvée Excelsior was introduced by Pierre-Marie's father in 2001, and it is fascinating to see how this wine has evolved over the years. The combination of long lees ageing (up to three years) and bottle age deliver sumptuous, golden, concentrated wines with luscious fruit ranging from confit lemon to dried apricot and mango. They age beautifully. The flagship cuvée is L d'Or, from vines planted in 1984, on a vein of granite in the village of Vallet. This is a hand-picked and hand-made wine that ages on its lees for almost a year in traditional glass-lined tanks and amphorae. Full and concentrated, this too is one for the cellar. As Marie says, 'We have the advantage of being able to make wine that can be drunk in the future.'

Famille Lieubeau

Château-Thébaud
www.lieubeau.com

The Lieubeau family farm a substantial 65 hectares of land in and around Château-Thébaud. François, Vincent and Marie are proud third-generation family wine growers. Their parents, Pierre and Chantal, are still very much present, too, their house virtually adjoining the winery. Theirs is a romantic history: Pierre's father was found as a baby abandoned outside the main hospital of Nantes and given the name Joseph Grégoire Lieubeau, the surname meaning 'beautiful place'. When he grew up Lieubeau settled in the Muscadet region, where he farmed many crops, including grapes, and passed the business on to his son. In the 1980s Pierre and Chantal decided to focus their attention solely on wine. Since then, they have built up a significant portfolio of vineyards and wines, including land in several *crus*: Château-Thébaud, Clisson and Goulaine. François Lieubeau shows me the family's flagship vineyard, a sloping 12-hectare plot close to the winery in Château-Thébaud. Here, the oldest vines date back to 1945 and are the source of the domaine's Château-Thébaud *cru* wine.

The vines are farmed organically. The *cru* wines are hand-picked, fermented with native yeasts and matured for at least two years on lees. Of these wines the Château-Thébaud is perhaps the most impressive, with an intensely mineral palate and a vein of savoury acidity running through it. It typically spends longer on lees (up to three years) compared to the other *crus*. The old vines that produce this wine grow on very thin topsoil over gneiss rock, a difficult medium for roots to penetrate. For the Clisson wine the vines are slightly younger, averaging 50 years of age. Here the soil is granite, but with up to two metres of sand on top. François explains that the combination of the soil and a more southerly site in Clisson will always give a riper style of wine, perhaps with a touch more alcohol. The Clisson wine tastes altogether softer and rounder, the mineral notes more in the background. Goulaine is the most recent addition to the Lieubeau *crus*, the first vintage being 2017. This is an even warmer site, and the first to ripen, with slopes facing from north-east to south-west. François describes the vineyard as 'a hill of sand with pieces of serpentinite in it'.[1] The Goulaine vine-

1 As an aside, geologist Alex Maltman, author of *Vineyards, Rocks and Soils*, notes that it is very unusual to find vines (or indeed any other plants) growing anywhere close to serpentinite, as its mineral composition can lead to toxicity, or problems with plant nutrient uptake.

yards rise from the mild marshes of the nature reserve towards the Butte de la Roche, a favourite sunset viewing spot where visitors can get a 360-degree view of Nantes and the Muscadet vineyard from a mere 47 metres above sea level. François characterizes the Goulaine wines as 'smooth and elegant'. I agree.

While the *crus* are the undoubted and age-worthy stars of the Lieubeau portfolio, another major bottling is 'Confluent', which comes from younger vines of 10 to 30 years old and spends around 12 months on lees. Fragrant and salty, this is a more classic style of Muscadet for earlier drinking. A recent addition is a fun range inspired by Jules Verne, a son of Nantes. With colourful labels and names recalling the author's novels, these are bottled early with no lees contact. François says, 'The next bottling will be Voyage to the Centre of the Earth – so we can talk about geology without being boring.' Who wouldn't drink to that?

Frères Couillaud, Château de la Ragotière

Vallet

www.freres-couillaud.com

This significant domaine of 80 hectares is situated in Vallet, at the eastern edge of the Muscadet appellation. It is unusual in the region for its outward-looking and market-focused approach: at least three-quarters of their wine is exported and most of the vineyard is devoted to producing non-Muscadet wines: 35 hectares of Chardonnay and 13 hectares of Sauvignon Blanc grow alongside the original 25 hectares of Melon. 'We need to listen to the market,' says seventh-generation winemaker Amélie. Nevertheless, Muscadet remains at the heart of the production.

The family has long had its roots in viticulture, but their ownership of the property here in Vallet dates only from 1978, when Amélie's father and his two brothers decided their existing 8-hectare domaine in nearby Mouzillon was not large enough to support the three of them, so bought the 25-hectare Château de la Ragotière, an aristocratic winemaking property of 30 hectares dating back to the fourteenth century. The family now also owns a 40-hectare estate in the appellation, Château la Morinière. Amélie and her husband Vincent took over in 2006.

The Château de la Ragotière range of Muscadets is of solid quality, with touches of flair to be found in individual wines. Les Schistes is Muscadet in the classic style, with a crisp, lemony, frank palate. The fruit comes from young vines of less than 20 years old and spends around seven months on lees. It is fresh, salty and inexpensive. The Vieilles Vignes is altogether more interesting, the vines here ranging from 40 to 90 years old – the oldest planted in 1932. Yields are slightly lower at around 45 hectolitres per hectare, and the wine spends 14 months on lees. It has mouthwatering pear and apple fruit with concentrated salinity and a touch of smokiness. The Cuvée Amélie is also from older vines and has a similar élevage to Vieilles Vignes, but malolactic fermentation gives a creamy, buttery style reminiscent of Chardonnay. An alternative bottling is a wine aged on lees for ten months in two different containers. The sunny site, adjacent to a granite quarry, gives a powerful wine with unusually high alcohol for a Muscadet. The first wine, Granite, was aged in the traditional subterranean, glass-lined, square, concrete vats. The second, La Petite Etoile Amélie et Vincent, was aged in 10-hectolitre ceramic eggs. According to Vincent, the eggs produce wine that is more evolved, richer and ready to drink.

Growing Chardonnay can be considered controversial in this monovarietal region, but the Domaine de la Ragotière Chardonnay 'Coeur de Schistes' is worth a mention. Modelled on the Chablis style it is a creamy and gently oaked wine that enjoys considerable success in North America in particular. The domaine devotes around 6 hectares to plantings of other varieties, including Viognier, Petit Manseng, Sauvignon Gris, Chenin Blanc and Pinot Gris.

Jérémie Huchet, Domaine de la Chauvinière

Château-Thébaud

www.jeremie-huchet-vigneron.fr

The bowler hat logo on the sign outside Domaine Jérémie Huchet seems slightly at odds with the casual French vigneron who saunters out to greet me, but the confusion is quickly dispelled: in French the name of the hat is *chapeau melon*, and Melon is, of course, the grape of Muscadet. Thankfully this branding is the full extent of formalities here, and there's not a rolled umbrella in sight. Huchet is a smiley father-of-three who lets his children design his wine labels as a summer holiday amusement. Despite his relaxed demeanour, however, his is a domaine to be reckoned with, and one of the most important quality estates of Muscadet.

Huchet is a fourth-generation winemaker who can trace his ancestry here back to 1680. It was his great-grandfather, a wine grower and miller, who purchased Domaine de la Chauvinière in Château-Thébaud where the winery is based today. His grandfather took over the 6 hectares of vines in 1945 and wine growing became more of a focus for the family. From the 1970s Jérémie's father began bottling the wines at the domaine and exporting. Jérémie, now at the helm, joined his father in 2001. Since this time the land under vine has expanded with the purchase in 2001 of 11 hectares of vines in Goulaine (Clos Les Montys), 13 hectares in Monnières-Saint-Fiacre (Château de la Bretesche) in 2006, and the 12-hectare Château de la Templerie, in the Clisson *cru*, in 2010.

The domaine produces wines under a range of labels reflecting the individual sites and, especially, the varied underlying geology. The Domaine de la Chauvinière *sur lie* bottling, from grapes grown on the granite plateau around the winery, is as good a classic expression of Muscadet as can be found anywhere: spritzy, salty, bright and smooth-textured. Also from Domaine de la Chauvinière comes the Chemin des Prières, a broader wine that has more intensity and something of a liquorice character, but with nothing overdone. 'It's typical Château-Thébaud,' says Jérémie, 'powerful but fine.' The wines from his Goulaine site, Les Montys, are fabulous. In particular, the Clos Les Montys Vigne de 1914 cuvée shows the sheer class and extraordinary balance of fruit that can be obtained from centenarian vines. It has a silky-smooth palate with enveloping richness and impressive, firm acidity. Overall, the wines have a distinctive house style. They are tight, linear and precise, leaning towards austerity at times. Jérémie aims to avoid what he calls 'bodybuilding wines' and is pleased that there has been a return to appreciation of wines in the lighter style he prefers. He also makes notable *cru* wines from Clisson, Château-Thébaud, Gorges and Monnières-Saint-Fiacre.

Jérôme Bretaudeau, Domaine de Bellevue

Getigné

There's nothing conventional about Jérôme Bretaudeau. Operating from a warehouse on the outskirts of the Muscadet appellation in Getigné, he is a self-taught winemaker who does exactly as he pleases, experimenting joyfully with the fruits of his vineyards to produce distinctive wines that have garnered a cult following. Bretaudeau is not as shy as his absence of website or social media might suggest. In the flesh he is an ebullient character, eager to share all manner of wines, from his latest bottlings to his potions bubbling left and right in barrels, concrete eggs or amphorae. The fact is, he does not need publicity; the world comes knocking on his unmarked door and the wines appear to sell themselves.

Jérôme is not from a winemaking family, but his passion for wine began when he was 16. 'The other kids were buying cigarettes; I bought wine,' he says, adding that his penchant was for Bordeaux – 'until it became unaffordable'. He worked for 13 years in viticulture before managing to set up on his own in 2005. He now farms 17 hectares, two-thirds of which are owned by him and one-third rented. The land is all in Muscadet communes, including Clisson, Gorges and Vallet, but Jérôme is clearly more of a freewheeler than a follower of rules. He grows nine different grape varieties, which are the base ingredients for his winemaking wizardry.

The approach is natural here. Hand-harvested grapes undergo spontaneous fermentation, and Jérôme avoids using any additives in the cellar. All of the wines go through malolactic fermentation, a stylistic choice. 'Everyone is afraid of malo here,' he says. His Muscadets are intense wines, the result of long fermentations and ageing. Muscadet Granit Clos des Perrières ferments for up to a year before further ageing in underground vats. It is an exciting wine, smoky, stony, rich and powerful, with mouthwatering acidity. He also makes a Muscadet from gabbro terroir. His best-known cuvée, Gaïa, from a small parcel of old vines, is a wine of concentration, salinity and mineral power. Long and rich, it ages for two years in a concrete egg. Jérôme has no interest in classifying these wines as *crus*. 'My standards are stricter,' he says, simply, and it's hard to see how *cru* status would benefit him: the wines speak for themselves.

But Muscadet is just the beginning of his creativity. Another white, La Justice, is a wonderfully energetic and complex blend of Chardonnay and Savagnin that ages in Slovenian oak. He makes an orange wine, Macération, from Pinot Gris grapes macerated for ten days and aged in amphora and barrel. It is bright orange, with an enticing floral perfume. Then there is Ornaté, an improbably ripe and supple wine from Merlot, and Statera, an astonishingly perfumed and succulent Pinot Noir. It's hard to believe either of these grew in the vineyards of Muscadet. Jérôme puts his nose into a glass of Statera, and his face lights up. 'It's like fireworks,' he says.

Other producers to try

Château de Cléray; Gadais Père et Fils (especially Plantation 1947 Monnières-Saint-Fiacre); Chéreau-Carré (good value *cru* wines); Apolline and Julien Braud (precise and pristine wines); Bonnet-Huteau; Bernard et Benoit Landron; Romain Moreau; Bruno Cormerais; Domaine de l'Ecu; Château du Coing; Michel Delhommeau; Domaine de la Bronnière.

COTEAUX D'ANCENIS

The 151 hectares of vines planted in the Coteaux d'Ancenis appellation are found in small parcels in the picturesque rural communes on either side of the Loire, centred around the town of Ancenis. Close to the river, the land is uncultivated, given over to cows, horses, orchards and gardens. The vines grow on slightly higher land on gentle, open slopes.

There are 40 producers here, of whom 10 are organic. The vineyards of Coteaux d'Ancenis intermingle with the Muscadet Coteaux de la Loire appellation of similar size, and among these Melon vines are pockets designated Muscadet Cru Champtoceaux. The climate is warmer with lower rainfall than the Muscadet vineyards to the south. The majority of the AOC Coteaux d'Ancenis production is dedicated to a lightly sweet wine from Pinot Gris, locally known as Malvoisie. This style, which accounts for two-thirds of the appellation's plantings, must contain between 20 and 40 grams residual sugar. It is popular locally as an aperitif but little known elsewhere. When I asked producer Benoit Landron where he sold this wine, he answered: 'Ancenis'. It is popular locally as an aperitif but doesn't travel much further than this.

Of greater interest are the rosés and reds that are grown in the rest of the appellation. They are typically 100 per cent Gamay, but growers have the possibility of adding Cabernet Franc as an accessory grape for up to 10 per cent of a blend. The rosés are always dry (useful to know in a region where rosé is often not) and can be no more than 12 per cent alcohol by volume. Grown on schist, these wines offer uncomplicated fruity enjoyment, usually at a bargain price. The reds are refreshing and light (typically 12.5% abv), and there are some inexpensive, easy-drinking gems here. Perfect summer picnic wines, they have soft, rich, raspberry, plum and cherry fruit, which may be accentuated by carbonic maceration before fermentation. The best examples have lip-smacking vibrancy and energy, and are very easy to enjoy.

Coteaux d'Ancenis – key facts

- **AOC granted**: 1973
- **Hectares in production**: 151
- **Annual production volume**: 6,900 hectolitres/920,000 bottles
- **Grapes**: whites from Pinot Gris (Malvoisie), rosé and red from Gamay Noir (with up to 10 per cent Cabernet Franc)
- **Styles**: 60 per cent sweet white, 29 per cent red, 11 per cent rosé
- **Soils**: mica schists and greenstone

Notable producers

Domaine des Galloires

Drain

www.galloires.com

At the crossroads between Anjou and the Nantais, this reliably good-quality domaine is run by the Toublanc family. It is an enterprise that goes back seven generations. Here, they farm 50 hectares and 50 cows. The domaine has always been a polycultural enterprise. Coteaux d'Ancenis reds are soft and juicy, made from 100 per cent Gamay. A wide range of other wines is produced, including several Anjou reds and a good crémant (under the Leon & Jean label). The family welcomes wine tourists for walks in the vines, tastings and gastronomic picnics.

Other producers to try

Domaine Merceron-Martin; Renou Frères & Fils; Domaine des Clérambaults.

FIEFS VENDÉENS

Wine grape growing in the far-flung Fiefs Vendéens can be traced back as far as the Roman era, but it developed significantly during the Middle Ages. Records from the eleventh and twelfth centuries show traders took advantage of the proximity of the vineyards to the coast and enjoyed brisk business shipping wines to Holland and England. The continued success of Vendée wines can be seen by the fact that, by 1880, 18,000 hectares were planted.

Fiefs Vendéens – key facts

- **AOC granted**: 2011
- **Hectares in production**: 356
- **Annual production volume**: 14,500 hectolitres/1.9 million bottles
- **Grapes**: white – Chenin Blanc, Chardonnay, Grolleau Gris, Sauvignon Blanc; rosé/red – Cabernet Franc, Négrette, Pinot Noir, Cabernet Sauvignon, Gamay, Grolleau Gris
- **Styles**: blended still wines, 42 per cent rosé, 36 per cent red, 22 per cent white
- **Soils**: mostly schist, gneiss, limestone, covered with aeolian sand or quartz

However, as in so many places, phylloxera arrived soon after, and nothing would ever be the same again. In response to the collapse of their livelihoods, one of the solutions tried by the desperate growers was to plant hybrid vine varieties which, it had been discovered, were resistant to the ravages of the insect. This practice was widely adopted in many parts of France, and nowhere more enthusiastically than here. By 1934, the Vendée was the number one producer of wines from hybrid varieties in the country. But this proved to be another blow for the newly recovering wine industry. Concerned that wines from these varieties were of far lower quality, and potentially even injurious to health compared to those made from the fruit of traditional *vinifera* vines, the government introduced a national ban on their production, requiring yet another new beginning for the poor wine growers of the Vendée.

It was only after the Second World War that attempts were made officially to revive the wine industry, which progressed to the now-obsolete VDQS (Vin Délimité de Qualité Supérieur) status in 1984. Improvements in the quality of the wines led to granting of AOC status in 2011. Today, this most westerly appellation of the Loire comprises little more than a dozen producers, farming fewer than 400 hectares across five distinct islands of vines: Brem-sur-Mer, Vix, Chantonnay, Mareuil and Pissotte. Leading wine grower Thierry Michon, of Domaine Saint Nicolas, recounts that a hundred years ago there were more than 300 vignerons in Brem alone.

Each of the geographical denominations is subject to slightly different regulations regarding the varieties that can be used, but all wines must be blends to have the right to the appellation. Whites must be Chenin-based, with a proportion of Chardonnay and the option to add Grolleau Gris or Sauvignon. Rosés are blends of Pinot Noir and Gamay. These are easy-drinking, light wines with a ready local market of coastal holidaymakers. Reds are based on Pinot Noir or Cabernet Franc, with compulsory inclusion of Négrette, a grape that gives a savoury edge to these early-drinking, fruity wines.

The vineyards are situated in the southern half of the Vendée, some distance south of the Loire. The climate is notably sunny here, with lower rainfall than the next closest Loire vineyard region of Muscadet. These are generally light, attractive wines. The two producers recommended

The irrepressible Thierry Michon, of Domaine Saint Nicolas

here are worth seeking out. Both are organic and each produces a wide range of wines, some of which fall outside the appellation. Jérémie Mourat runs a sophisticated business that encompasses a well-developed oenotourism side with vineyard tours and a shop. His vineyards account for close to half of the land in the appellation. Thierry Michon, owner of Domaine Saint Nicolas, is a creative soul with a highly developed sense of fun. A keen accordion player, he has been known to greet his guests with a serenade on arrival. Another of his proclivities is performing headstands, evidenced by his many photos that document him upside-down in locations throughout the world. He and his sons produce immensely likeable and interesting wines, often from a single variety such as Chenin Blanc or Pinot Noir, which are therefore not entitled to the appellation. Like so many Vin de France producers, exerting his creative freedom is far more important to Thierry than sticking to a prescribed formula.

Notable producers

Domaine Saint Nicolas, L'Ile d'Olonne: www.domainesaintnicolas.com; J. Mourat, Mareuil-sur-Lay-Dissais: www.vignobles-mourat.fr.

The Italianate town of Clisson

When in the Nantais

Nantes is the city of Muscadet, less than half an hour by car from the vineyards. It is a lively university city with plentiful dining options and opportunities to taste a great selection of Muscadets. For a special meal overlooking the Loire (and right next door to the fascinating Jules Verne museum), the Michelin-starred L'Atlantide 1874 Maison Guého (www.atlantide1874.fr) offers a serene and fresh dining experience with a wine list that features aged Muscadet *cru* wines. In the heart of the vineyards of Muscadet, there are few dining options, but the Auberge La Gaillotière (www.auberge-la-gaillotiere.fr) in Château-Thébaud is a perfect lunch spot, offering a range of Muscadets by the glass to accompany the tasty local food. Close by is the unmissable Porte-Vue in the village of Château-Thébaud, a spectacular glass-floored steel platform that juts out over the river Maine, 30 metres below. It offers not only a dizzying view of the river, but also the vineyards and the rocks beneath them. In the summer a wine bar offers Muscadet tastings here. The Italian-styled town of Clisson is an unexpected delight, centred on the river Sèvre. Usually a quiet and charming town with a medieval castle and winding streets, it is transformed every June when it hosts Hellfest, one of the largest metal music festivals in Europe.

The Porte-Vue of Château-Thébaud with vines in the background

5

ANJOU: SOMETHING FOR EVERYONE

Anjou means different things to different people. It is a contradictory place, simultaneously perhaps the most and least interesting region of the Loire Valley. Here, the most refined and sublime wines can be found, including world-class Chenins and the only real cult red wine of the Loire, Clos Rougeard. But alongside such vinous highlights are many uninspiring wines that have, over the years, eroded the region's historic reputation for quality. But today, there is much to be both hopeful and excited about from the artisanal producers in the vineyards and villages of Anjou, whether from historic domaines or newly created enterprises. In recent years, some notable newcomers have moved into the area, taking advantage of Anjou's under-the-radar status and inexpensive land. New terroirs are being discovered (or rediscovered) and there is now growing interest in the truly excellent wines that are coming out of the region. Although there are still many wines of limited interest in Anjou, in a way this makes unearthing its treasures a more rewarding pursuit. And today there are plenty of jewels to be found in the cellars of Anjou.

A third of all the wine produced in the Loire Valley comes from this large area, which covers land in 151 communes across three departments: Maine-et-Loire, Deux-Sèvres and Vienne. Most of the vineyards are to the south of the Loire, following the broad sweep of the river from the edges of the Nantes conurbation eastwards to Saumur. The most concentrated section of wine growing is between the riverside towns of Angers and Saumur. It is an area that spans a

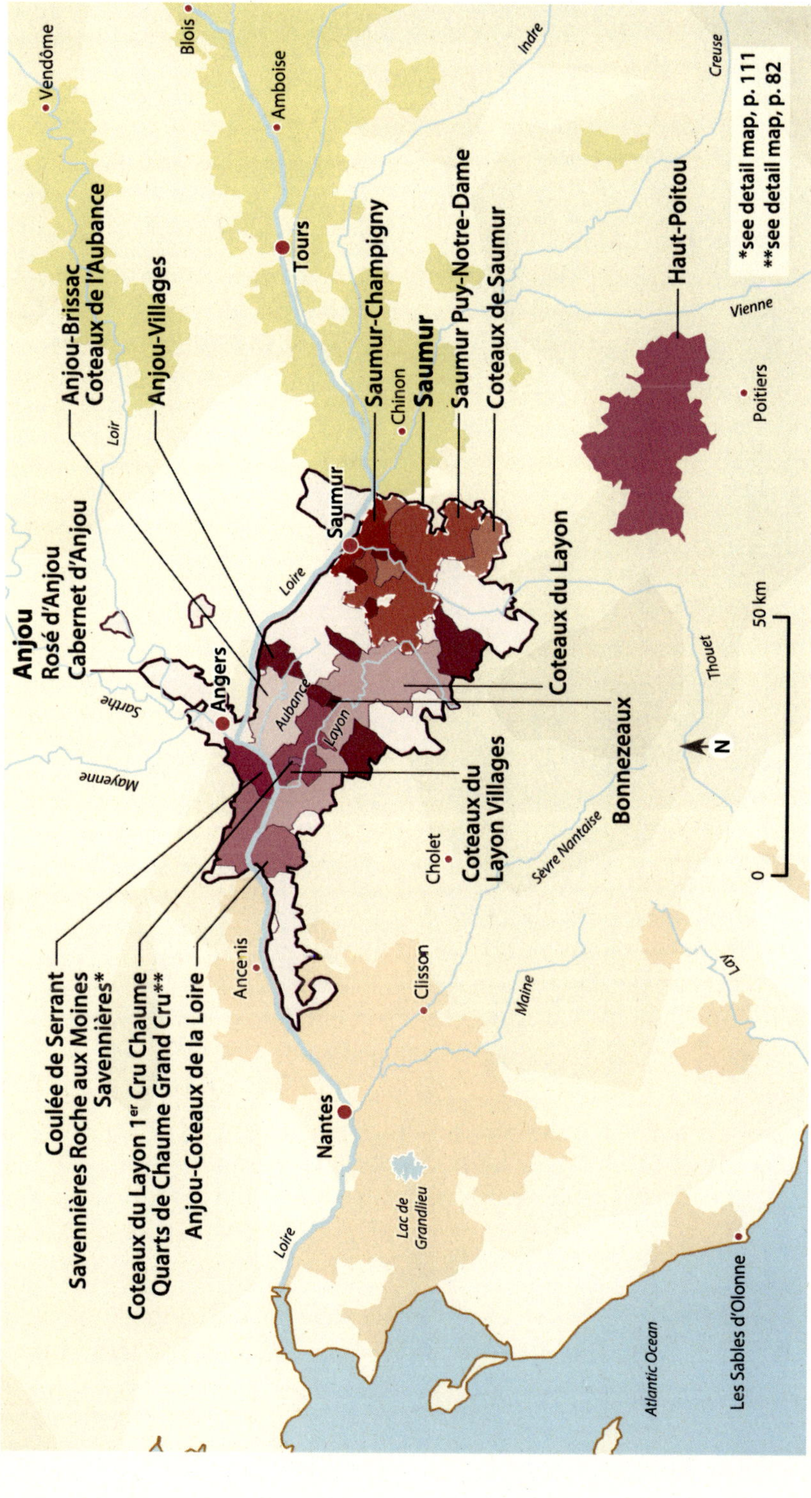

Map 5: Anjou, Saumur and Haut-Poitou

wide range of terroirs, multiple smaller rivers and two distinct geological eras.

Two of the Loire's tributaries feature prominently in the appellations of Anjou: the Layon and the Aubance. Both flow from the south to join the Loire close to Savennières. The Layon is geographically small but geologically significant, for its south-east to north-west course marks the line of the Layon fault. This is the frontier between the ancient volcanic and metamorphic rocks of the Armorican Massif to the west and the younger, limestone-based soils of the Paris Basin to the east. It marks the divide between the so-called Anjou noir (west) and Anjou blanc (east), named for the predominant dark or light colour of the rocks and soils. This divide splits the region more or less down the middle. In places, the landscape reflects the drama of the earth's past, with gentler slopes on the left bank of the river Layon and steep, rocky inclines to the right. The geology here is, as many a winemaker will tell you, 'chaotic' – but it is precisely this chaos which allows creative winemakers to produce a seemingly endless selection of cuvées to reveal in their wines the multi-faceted expressions of these diverse soils.

Today you can find every Loire wine style in Anjou, and many producers make a wide range of different wines, even though some are clearly better adapted to their terroir than others. Before phylloxera, the picture was quite different: Chenin Blanc was by far the dominant grape, and it was especially valued for its ability to produce sweet wines in an era where sugar and sweetness were hard to come by. Post-phylloxera the quality hillside vineyards were abandoned in favour of sites that were easier to work to facilitate mass-market, high-volume production. This was sent largely to the western parts of France where wine production was not possible. The reputation of Anjou diminished over the course of the century following phylloxera until it became little more than a distant memory. The goals of fulfilling demand and following fashion led to the demise of the region's signature Chenin Blanc in favour of reds, often of lower quality varieties, or varieties generally unsuited to this northerly climate (Cabernet Sauvignon, for example). A good way to sell the wines from these tart grapes turned out to be converting them into rosés, with the acidity softened by a good dose of sweetness. Even today, most producers make a wide range of styles of wine, and it is not uncommon to find vignerons turning their hands to reds and rosés as well as Chenin in all its guises.

Anjou has more appellations than any other Loire region, and the most varied output. Cabernet d'Anjou is the single largest appellation of the Loire, with an average production of 45 million bottles.

The Anjou appellations

- **The Anjous**: Anjou; Anjou Villages; Anjou Brissac; Cabernet d'Anjou; Rosé d'Anjou
- **The sweet Chenins**: Quarts de Chaume Grand Cru; Coteaux du Layon Premier Cru Chaume; Coteaux du Layon; Coteaux de l'Aubance; Bonnezeaux; Anjou-Coteaux de la Loire; Coteaux de Saumur
- **The Savennières appellations**: Savennières; Savennières Roche-aux-Moines; Coulée de Serrant
- **Saumur**: Saumur; Saumur Mousseux/Fines Bulles; Saumur-Puy-Notre-Dame; Saumur-Champigny
- **Haut-Poitou**

Anjou – key facts

- **Hectares in production**: 18,800 (33 per cent of the Loire total)
- **Annual production volume**: 989,000 hectolitres (36 per cent of the Loire's output)
- **Styles**: 50 per cent rosé, 27 per cent sparkling (which includes all of Crémant de Loire, much of which is produced here but that also covers a wide area across Anjou and Touraine), 14 per cent red, 9 per cent white

Within France Anjou is still most readily associated with basic quality, off-dry and semi-sweet rosés made under the appellations Rosé d'Anjou and Cabernet d'Anjou. As these wines are little seen outside the country, it is easy to forget just how important they are to the local economy. Together, these two appellations account for well over half of Anjou's vinous output. What's more, Cabernet d'Anjou (the sweeter of the two styles) is the largest-volume appellation of the entire Loire Valley: one in ten Loire wines is a Cabernet d'Anjou. Producers trying to establish Anjou as a quality wine-producing region might be exasperated that this is the most common association some people have with Anjou, but others celebrate its undeniable commercial success. It has even been suggested that these rosés could be marketed as a viable alternative to coffee as an afternoon pick-me-up for tired mothers.

Some producers see little value in using the generic Anjou appellation and abandon the label altogether, preferring to bottle their wines as Vin de France. As in other parts of the Loire, Anjou has its fair share of winemakers who prefer to have the freedom to make wines in any way they wish and from any grapes they choose, which also leads them down the Vin de France route. But outside of France, Anjou is a bit of an unknown quantity, which may prove to be its trump card. The name carries no negative quality associations for the vast majority of wine drinkers around the world. For most of us, the new wave of high-quality wine that has emerged under the Anjou appellation is simply an exciting discovery.

An important category that falls within the Anjou region is sparkling wine production, which centres around the town of Saumur. Jean Baptiste Ackerman developed the sparkling industry here in the early nineteenth century. After spending some time in Champagne, he came to work in Saumur and quickly realized that here was an ideal environment for producing and maturing sparkling wines similar to those in Champagne. His success led other companies to follow suit. Today Saumur is the most important centre for sparkling wine production in the Loire, and the Ackerman business continues to thrive along the banks of the river in Saumur. Wines are made in the traditional method and age for at least nine months on lees. The large houses produce a wide range of wines of consistent quality at a fair price. This is an increasingly important and growing market.

Rosé and sparkling dominate the quantity of wine produced in Anjou, but when it comes to quality it is the minority white wines that are gaining the most attention. Anjou is the almost certain birthplace of Chenin Blanc and the grape has been growing here continuously for centuries. The Anjou vineyards are the source of some of the world's most outstanding wines from this versatile variety. Anjou was historically best known for the quality of its sweet wines, since it has naturally favourable conditions for the late-ripening Chenin: sunny slopes, a relatively dry climate and autumn humidity from small rivers encouraging development of noble rot (*Botrytis cinerea*). Even if sweet wine is less fashionable than it once was, there are still no fewer than seven separate appellations for sweet wine here, including the Loire Valley's only *grand cru* vineyard, the 40-hectare Quarts de Chaume. (Chenin is also the main ingredient in Saumur sparkling wine.) Even though Chenin no longer dominates the Anjou landscape as it once did, for 12 of the region's appellations, it remains the only permitted variety growers can use.

While the sweet wines from Anjou can be superb, there is more interest today in quality dry wines from Chenin. Whether from the long-established bourgeois properties of Savennières or Saumur, or the plain, generic Anjou appellation, a new quality standard has been established for these wines, from newcomers bringing ideas that are influencing other producers, or from the younger generations of existing family growers. Often grown on terroirs better known for sweet wines, these dry expressions have immense character and interest. Depending on the soils where they are grown, they show power, finesse, minerality and longevity, demonstrating Chenin's ability to express myriad nuances of terroir. The winemakers of Anjou and Saumur have been working with the INAO on a long-term project to attempt to gain recognition for the best sites and to create *crus* within the appellations. In Anjou these currently number nine and in Saumur six, the latter including the locally renowned Brézé. If past INAO wine reforms are anything to go by, change will occur at a snail's pace, and at this early stage the growers still need to agree among themselves on what it is that they want. In Saumur, for example, the drawing of the boundaries of the different *crus* is causing some predictable upset, while for the Anjou *crus* one of the most contentious issues is that the INAO insists that *crus* be for dry styles of wine only. This is a problem for growers who wish to retain the option of making either sweet or dry wines, depending on the vintage conditions of the year (which is how the basic appellation system operates in Vouvray, for example). For now, the dry wines produced from sweet wine appellations are labelled simply as generic Anjou blanc (see box, p. 77).

Anjou reds centre on Cabernet Franc, and come in a range of styles. The best known are those grown on the warm tuffeau-based soils of Saumur-Champigny where most comprise 100 per cent Cabernet Franc. Elsewhere in Anjou, blended wines are more common. Cabernet Franc remains in the majority, usually blended with a little Cabernet Sauvignon. Some light, early-drinking wines are made under the appellation Anjou Gamay, which is usually made wholly from this juicy, red-fruited variety.

One of the beauties of Anjou is its variety, but it is nevertheless encouraging to see it increasingly playing to its natural strengths. Chenin Blanc is making a comeback as the grape of choice for new plantings, and even replacing red wine grapes. This is in recognition of its historic presence in the region, its adaptability to the soils and growing conditions of Anjou and a rise in demand for white wines worldwide.

Sparkling wines are following the global trend towards drier, more elegant styles, and red wines too are moving to a lighter and more restrained style that suits the terroir where they are grown.

THE ANJOUS AND SWEET WINE APPELLATIONS

The Anjous: Anjou, Anjou-Villages, Anjou Brissac, Rosé d'Anjou and Cabernet d'Anjou

The catch-all, generic appellation of Anjou covers the large area that encompasses all the appellations and all the soils of the Anjou region. Wines can be red, white, or sparkling (rosés have their own separate appellations). Production is around two-thirds red, and usually restricted to wines grown in the western part of Anjou. While vineyards further east could in theory be labelled as Anjou as well, they are entitled to the more highly regarded (or at least, more geographically specific) Saumur appellation (see p. 127). Compared to the Saumur designations Anjou affords more flexibility in terms of the grapes allowed in blends, and permitted yields are slightly higher.

Anjou reds can be made from Cabernet Franc, Cabernet Sauvignon, Pineau d'Aunis, Grolleau and Gamay. Basic Anjou red is a light wine, destined for early and, usually, local consumption. At 60 hectolitres per hectare, permitted yields for reds are a little more generous than for Anjou-Villages (55 hectolitres per hectare) and Anjou Brissac (50 hectolitres per hectare). The Anjou-Villages appellation is valid only for red wines and covers 43 of the region's 151 communes. Here, vines grow on schist-based soils either side of the river Layon and towards the Loire. With slightly lower yields and a requirement for the wines to be made only from Cabernet Franc and Cabernet Sauvignon grapes, they tend to be more structured than basic Anjou reds. Anjou Brissac, another red-only denomination, is the highest quality level of these three appellations for reds. It covers 10 communes, clustered around the Aubance river. Vines grow on gentle slopes of schist – the classic Anjou noir soils. With lower yields, higher planting density and a requirement to age the wines until the following summer before bottling, these are slightly more age-worthy wines and in ripe vintages can show attractively dense black fruit and firm structure.

Anjou whites must comprise a majority of Chenin Blanc, but the best examples are all 100 per cent Chenin. There is a wide range in quality and prices of Anjou blanc wines. Some are among the cheapest wines to be found in the Loire and others among the most expensive, selling for more than 100 euros per bottle. In between these two extremes are some very good wines that offer excellent value for money. Prices are rising as the world discovers these wines. Many of the best Anjou blancs come from historic Chenin sites around the Layon Valley and the hill of Chaume. This is the home of the top sweet wine terroirs of the Loire, appellations such as Coteaux du Layon, Coteaux du Layon Premier Cru Chaume and Quarts de Chaume Grand Cru.

The Anjou white wine revolution

When the sweet wine appellations were created, it was well known that it was not possible to produce sweet wines in every vintage: in some years conditions are simply more favourable than in others so, in the past, the wines were sometimes made in a drier style. When it was discovered that sugar could be added to underripe grapes to produce a sweet wine, the practice (known as chaptalization) was enthusiastically adopted by many of the growers in these appellations to fulfil the high demand at the time for sweet wines. This meant that it was possible for them to make sweet wine every year, albeit of lesser quality, because the sugar was coming from a bag rather than from the natural sweetness in the grapes. The chaptalized wines may have been of lesser quality, but for the growers, at least there was a viable and reliable business in making sweet wines in Anjou. However, as a consequence of this, the appellation rules were revised by the INAO to reflect the fact that the wines were now always sweet. There was also a reasoning from the INAO (which continues today) that it would make life easier for consumers, who might be confused by whether the wines they were buying from an appellation would be sweet or dry. In the 1990s, therefore, the regulations were changed to stipulate that wines from these vineyards had to be sweet to qualify for the appellation.

Making sweet wine is a high-risk endeavour in most places, and Anjou is no exception. It entails waiting until the end of the season and hoping that the weather will allow the grapes to continue to ripen and that, ideally, noble rot (*Botrytis cinerea*) will allow production of a small quantity of botrytized sweet wine. Many things can upset this plan. Foremost of these is the most obvious: the weather. Although the best sweet wine regions are located where conditions are likely to be favourable, very often the right conditions do not occur. Rain is always a risk at the end of the growing season. And when that happens, it's game over, potentially no wine at all, which is not a sustainable option for anyone economically (or, one has to imagine, emotionally). Even if conditions are good, the market for sweet wines is by no means assured in the twenty-first century. In a world where sugar and sweetness are all too plentiful for most of us these wines are not as obviously desirable as they used to be. So a grower wishing to make a natural, high-quality wine with no recourse to chaptalization is faced with a decision: wait for naturally favourable conditions to pick and risk losing the entire crop, or pick earlier and make some dry wine (or maybe make mostly dry wine instead). In a way, this is going back to the old traditions of winemaking: the sweet wine terroirs sometimes deliver and sometimes don't. But now that dry white wine is in greater demand than sweet wine, there has been a move towards making far more dry wine than sweet from these historic sweet wine sites in Anjou. This is the new face of high-quality Anjou blanc.

The first forays into dry whites came at the end of the 1990s. The middle years of the decade, 1995, 1996 and 1997, were all great vintages for sweet wines, but 1998 and 1999 were not. Now, with more than 20 years' experience in making successful dry wines, and an influx of new talent and investment, there is no doubt that the great sweet wine terroirs can make great dry wines as well.

The dry white wines in the sweet appellations are currently entitled only to use the catch-all generic appellation Anjou. The growers in Quarts de Chaume would like to be able to use their *grand cru* appellation for both sweet and dry wines, but so far this has not gained any traction with the INAO. Wheels are in motion, however, for recognition of a *premier cru* called Ronceray, which will be for dry wines made within the boundaries of the Quarts de Chaume Grand Cru and Coteaux du Layon Premier Cru Chaume. As a result, some wines are now labelled with the name Ronceray in anticipation that this will become recognized as a *premier cru* in the near future.

The Loire is the second-largest producer of pink wines in France after Provence. But unlike the recherché image of their southern counterparts the semi-sweet rosés of the Loire have no pretensions of being anything but inexpensive thirst-quenchers. Bottles (and boxes) of Rosé d'Anjou and Cabernet d'Anjou fill the shelves of the local supermarkets and prices are uniformly low. Cabernet d'Anjou is the only AOC in France for *rosé tendre*, and it is the biggest-selling rosé of any kind in French supermarkets. The wine must contain at least 10 grams per litre of residual sugar, but is often far sweeter, resembling an alcoholic version of a diluted soft fruit drink. Little seen outside France, much is consumed locally, where it is recommended as an everyday wine to consume with anything from charcuterie to macarons and fruit desserts. Rosé d'Anjou is a drier version of Cabernet d'Anjou that may contain a wide selection of varieties beyond the Cabernets, including Grolleaus Noir and Gris, Gamay and Côt. Permitted yields for Rosé d'Anjou are a substantial 65 hectolitres per hectare, slightly higher than the 60 hectolitres per hectare allowed for Cabernet d'Anjou. These off-dry wines, with a minimum of 7 grams of residual sugar, can be attractively fruity and vinous, with the light sweetness balancing the naturally crisp acidity of the grapes.

Anjou – key facts

- **AOC granted**: 1936, including AOC Anjou Gamay
- **Hectares in production**: 1,375
- **Production volume**: 62,200 hectolitres/8.3 million bottles
- **Styles**: 58 per cent red, 36 per cent white, 6 per cent sparkling
- **Grapes**: red – Cabernet Franc, Cabernet Sauvignon, Pineau d'Aunis (maximum 10 per cent), Grolleau Noir (maximum 20 per cent), Gamay; white – Chenin Blanc (80 per cent minimum), Sauvignon Blanc, Chardonnay; Anjou Fines Bulles sparkling – Chenin Blanc (minimum 70 per cent), Chardonnay (maximum 20 per cent), Cabernet Franc, Cabernet Sauvignon, Gamay, Grolleau, Pineau d'Aunis
- **Soils**: schist soils in Anjou noir, tuffeau limestone in Anjou blanc

Anjou-Villages – key facts

- **AOC granted**: 1987
- **Hectares in production**: 135
- **Production volume**: 5,800 hectolitres/770,000 bottles
- **Styles**: 100 per cent red
- **Grapes**: Cabernet Franc, Cabernet Sauvignon
- **Soils**: schists, some gravel

Anjou Brissac – key facts

- **AOC granted**: 1998
- **Hectares in production**: 66
- **Production volume**: 2,400 hectolitres/320,000 bottles
- **Styles**: 100 per cent red
- **Grapes**: Cabernet Franc, Cabernet Sauvignon
- **Soils**: schist, with quartz, volcanic rock, sand

Cabernet d'Anjou – key facts

- **AOC granted**: 1964
- **Hectares in production**: 6,200
- **Production volume**: 330,000 hectolitres/44 million bottles
- **Styles**: 100 per cent semi-sweet rosé (minimum 10 grams per litre residual sugar)
- **Grapes**: Cabernet Franc, Cabernet Sauvignon
- **Soils**: schist soils in Anjou noir, tuffeau limestone in Anjou blanc

Rosé d'Anjou – key facts

- **AOC granted**:1957
- **Hectares in production**: 1,860
- **Production volume**: 112,000 hectolitres/15 million bottles
- **Styles**: 100 per cent off-dry rosé (minimum 7 grams per litre residual sugar)
- **Grapes**: Grolleau, Cabernet Franc, Cabernet Sauvignon, Pineau d'Aunis, Gamay, Côt
- **Soils**: schist soils in Anjou noir, tuffeau limestone in Anjou blanc

The great sweet Chenins of Anjou

The sweet wines of Anjou have had their ups and downs over the last 50 years. Changing fashions and changing winemaking practices in the second half of the twentieth century resulted in a reduction in the quality of the wines being made, often from higher yields than desirable and with significant levels of chaptalization and added sulphur. After the ravages of phylloxera and then the decimation of the workforce and morale caused by the Second World War, many of these vineyards were pulled up or abandoned. Furthermore, the 1960s, 1970s and 1980s saw some of the worst vintages of the century, which discouraged winegrowers further and did nothing to improve the quality of the wines. Between 1975 and 1985 the Coteaux du Layon vineyards shrank by 40 per cent. It was not until the extraordinary twin vintages of 1989 and 1990 that things began to look up for growers in the Layon Valley, as they were able to make large quantities of excellent wines in conditions they had never witnessed before. Since this time, there has been a greater focus on quality over quantity, by limiting yields, harvesting by hand and improving winemaking practices.

The Loire's sweet wine appellations are mainly found around the Layon river, a meandering, narrow tributary of the Loire to the south of Angers. Records show that wines from Chenin grapes grown along this 50-kilometre stretch of river have been valued for their lush, sweet wines since at least the sixteenth century. In some ways, little has changed. Chenin is still the hero grape for the Loire's sweet wines, and the only permitted variety in these appellations. The landscape features steep, narrow valleys and broader, shallower slopes, all benefiting from low rainfall during the growing season compared to the rest of the region. The vines also enjoy good exposure and ventilation, both of which naturally favour early ripening and the ability to produce late-ripening styles (even though, as noted above, this is not without risk). In places (such as Domaine Belargus's Treilles vineyard, see box, p. 94), Mediterranean plants and fauna thrive happily alongside the local vegetation, proof of the especially warm and dry nature of this microclimate. The proximity of the vines to the Layon is highly favourable to the development of botrytis. It is hard to believe that this trickle of a stream is responsible for the magical transformation of Chenins grown here, but its presence is never far away. In Chaume, the river winds around the foot of the hill, mirroring the contours of the slopes. On cool, damp autumn mornings a mist rises from the river to shroud the ripe grapes with moisture, allowing the spores of *Botrytis cinerea* fungus to penetrate the grapes. As the moisture evaporates later in the day, warm and sunny afternoons encourage *pourriture noble* to take hold in the grapes.

The most prestigious sweet appellation of Anjou, and indeed the Loire Valley, is the Quarts de Chaume Grand Cru, which achieved its

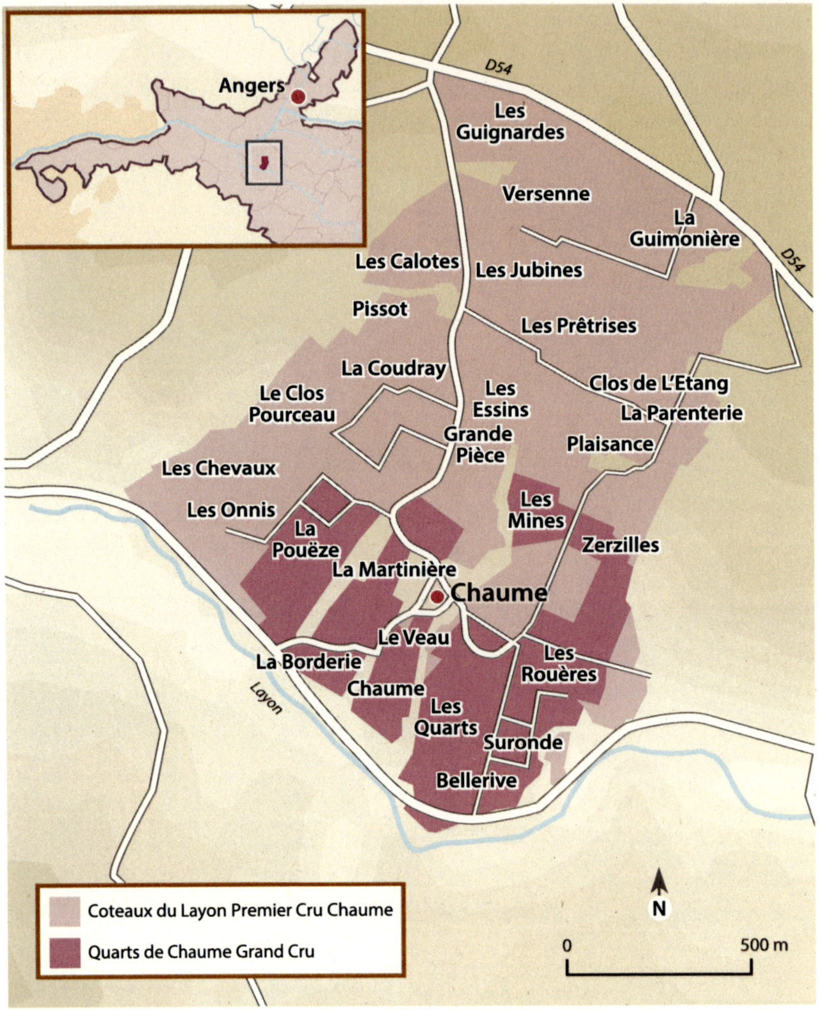

Map 6: Quarts de Chaume Grand Cru and Coteaux du Layon Premier Cru Chaume appellations

lofty status in 2011. The name of the site dates back to the fifteenth century, when the vineyard was owned by the Abbey of Ronceray, a female monastic order in nearby Rochefort. The nuns leased the vines to tenants who paid them with the best quarter ('*quart*') of their harvest, keeping the remaining three *quarts* for themselves. The vineyard was originally classified as an appellation in 1954, and the 2011 update details the conditions needed in order to produce wines worthy of the status of *grand cru*. The rules stipulate tiny yields of 20 hectolitres per hectare. All grapes must be harvested by hand and in several passes (which is, in any case, a practical necessity when picking overripe or botrytized fruit). The natural sweetness before fermentation is at a minimum of 298 grams per litre of sugar, which equates to a potential 18% abv. The minimum level of sweetness after fermentation must be 85 grams per litre – all natural of course, chaptalization is expressly forbidden – and wines are often well in excess of this level (the Domaine Belargus Quarts de Chaume 2021 has over 200 grams of residual sugar per litre, for example).

The Loire's top vineyards: Quarts de Chaume and Chaume

The 40-hectare site of Quarts de Chaume falls within the single commune of Rochefort-sur-Loire. What makes it so special? It enjoys the perfect combination of location, exposure (to the east, south and west), and a complex, unique mixture of soils including Brioverian schists, puddingstone, siliceous rocks and volcanic spilite. The clay topsoils are very thin – to the extent that in places the bedrock can be seen. These soils have little water-holding capacity and the sloping site ensures easy drainage. To the west, the Mauges hills provide a Foehn effect, limiting rainfall and protecting the vines from north and west winds. The Quarts de Chaume Grand Cru occupies the lower portion of the hill of Chaume that rises abruptly from the River Layon. The top of the hill is the highest point for miles around – albeit only 100 metres above sea level – and the vines grow at altitudes of between 25 and 75 metres. The peculiarities of this site reinforce the early ripening of the grapes, their susceptibility to botrytis and the consequent ability of Quarts de Chaume wines to express this unique terroir. Simply put, nowhere else is quite this perfect.

The second high-quality appellation on the hill of Chaume is the Coteaux du Layon Premier Cru Chaume, the Loire's only designated *premier cru*. While these plots enjoy the same geological complexity of soils as the Quarts de Chaume vineyard, they are more exposed, especially at the higher altitudes where the Foehn effect of the Mauges is less obvious and the vineyard more open to the elements. The Chaume vineyard surrounds that of the more protected Quarts de Chaume. In the *premier cru*, ripening is somewhat less precocious than in Quarts de Chaume and the conditions lead to more shrivelled grapes, which are slightly less susceptible to botrytis. In size, the area is double that of Quarts de Chaume, at around 80 hectares. Yields are limited to 25 hectolitres per hectare. Otherwise, conditions of production are very similar for both.

Quarts de Chaume Grand Cru – key facts

- **AOC granted**: 1954, designated *grand cru* in 2011
- **Hectares in production**: 17
- **Production volume**: 130 hectolitres/17,000 bottles
- **Styles**: 100 per cent sweet white
- **Grapes**: Chenin Blanc
- **Soils**: complex schists and puddingstones

Coteaux du Layon Premier Cru Chaume – key facts

- **AOC granted**: 1950, designated *premier cru* 2011
- **Hectares in production**: 29
- **Production volume**: 500 hectolitres/67,000 bottles
- **Styles**: 100 per cent sweet white
- **Grapes**: Chenin Blanc
- **Soils**: schist terraces on slopes of Chaume, shallow soils

The largest sweet wine area, Coteaux du Layon, extends across 13 communes that cover the full course of the Layon from Cléré-sur-Layon until it flows into the Loire at Chalonnes-sur-Loire some 50 kilometres downstream. There have been vines grown along both sides of the river for as long as anyone can remember. The soils are complex and varied, and are often quite different from one side of the river to the other because the valley of the river marks a major geological fault in the land (the Layon fault). On the left (south) bank the land descends in gentle slopes to the river. These are ancient Precambrian schists and sandstones. The right bank, however, rises very steeply, sometimes vertiginously, making the land very difficult to work. These are younger, Paleozoic schist soils, with veins of spilite, phtanite and, in places, a bedrock of coal, which was mined here up until the twentieth century. After phylloxera destroyed the vineyard, many of these difficult slopes were abandoned in favour of easier places to work, especially with the advent of mechanization.

There is a separate appellation for the villages in the heart of the growing area of the Layon, close to the hill of Chaume, which may add their name to the basic Coteaux du Layon appellation. These are: Beaulieu-sur-Layon (or 'Beaulieu'), Faye d'Anjou ('Faye'), Rablay-sur-Layon ('Rablay'), Rochefort-sur-Loire ('Rochefort'), Saint-Aubin-de-Luigné

('Saint-Aubin') and Saint-Lambert-du-Lattay ('Saint Lambert'). These villages benefit from more favourable growing conditions than the villages further upstream and can produce wines with higher levels of natural sweetness. One notable exception to this, however, is Bonnezeaux, a small appellation of some 80 hectares situated further upstream on the right bank of the Layon in the commune of Thouarcé. Here three exposed, south-west-facing steep slopes cover a band of land that is 2,800 metres long and 500 metres wide. The climate is warm and dry here, with around 100 millimetres less rain than the average for the Anjou region. The soils are complex and similar to those further downstream, with particularly shallow topsoil covering sometimes-exposed schist bedrock that drains easily and is readily warmed by the sun. Proximity to the Layon favours development of botrytis in autumn. These growing conditions are perfect for producing late-harvest wines, and those of Bonnezeaux can be long-lived. The prestige of this site for wine growing has long been recognized. The minimum sweetness level is 51 grams per litre, but many of the wines are far sweeter than this.

Coteaux du Layon Villages – key facts

- **AOC granted**: 1955
- **Hectares in production**: 94
- **Production volume**: 1,700 hectolitres/228,000 bottles
- **Styles**: 100 per cent sweet white
- **Grapes**: Chenin Blanc
- **Soils**: schists, sandstone, sandy gravels
- **Villages**: Beaulieu-du-Layon, Faye d'Anjou, Rablay-sur-Layon, Rochefort-sur-Loire, Saint-Aubin-de-Luigné, Saint-Lambert-du-Lattay

Coteaux du Layon – key facts

- **AOC granted**: 1950
- **Hectares in production**: 1,204
- **Production volume**: 34,000 hectolitres/4.5 million bottles
- **Styles**: 100 per cent sweet white
- **Grapes**: Chenin Blanc
- **Soils**: schists, sandstone, sandy gravels

Bonnezeaux – key facts

- **AOC granted**: 1951
- **Hectares in production**: 43
- **Production volume**: 800 hectolitres/107,000 bottles
- **Styles**: 100 per cent sweet white
- **Grapes**: Chenin Blanc
- **Soils**: steep slopes of schist and sandstone with quartz and phtanite

Unlike the Layon Valley's formation by violent geological upheaval, the slopes of the Coteaux de l'Aubance were sculpted by slow weathering, resulting in low, rounded hills instead of steep inclines. The Aubance is a minor tributary of the Loire that runs between the Layon to the south and the Loire to the north, and parallel to both. Seven communes either side of the river have the right to produce wines under the name of this appellation. The area is relatively small, with little more than 100 hectares in production. Winemakers in the vicinity produce the red wines of the Anjou Brissac appellation as well as sweet Chenins. Less well known than the larger Coteaux du Layon, the Coteaux de l'Aubance suffered similar dips in quality in the decades following the Second World War, but recent years have seen winegrowers keen to improve the standing of their appellation, producing better quality grapes and choosing Chenin clones with smaller berries that are better adapted to producing late-harvest wines. The wines are of similar sweetness levels to Coteaux du Layon, and similar in style.

Coteaux de l'Aubance – key facts

- **AOC granted**: 1950
- **Hectares in production**: 146
- **Production volume**: 3,800 hectolitres/507,000 bottles
- **Styles**: 100 per cent sweet white
- **Grapes**: Chenin Blanc
- **Soils**: shallow schists

The final two sweet Chenin appellations of Anjou are the tiny Anjou-Coteaux de la Loire and Coteaux de Saumur, situated at opposite edges of the Anjou region. Anjou-Coteaux de la Loire consists of a few small slopes with shallow, primary soils of the Armorican Massif in the far west of Anjou, immediately to the west of Angers. Never more than 3 kilometres from the Loire, these vineyards are heavily influenced by its presence. The climate is oceanic and, as for the other appellations in the vicinity, the vines are sheltered from the worst excesses of rainfall and wind by the hills of the Mauges to the west. The river also affects the climate of the appellation. To the north of the river, the vines enjoy southerly exposure; to the south, they benefit from the river draining cold air to warm the north-facing slopes. And of course, the closeness of the river assists in creating ideal conditions for botrytis to take hold. Nevertheless, the production is very small, amounting to an area that on average is a mere 15 hectares annually. Producers here make more sparkling wine and still, dry Anjou blanc, but the sweet wines are valued for their high ripeness without excessive alcohol, due to the poor soils; when Chenin's natural vigour is contained, it is better adapted to produce wines of this style.

Anjou-Coteaux de la Loire – key facts

- **AOC granted**: 1946
- **Hectares in production**: 15
- **Production volume**: 390 hectolitres/52,000 bottles
- **Styles**: 100 per cent sweet white
- **Grapes**: Chenin Blanc
- **Soils**: shallow schist

Coteaux de Saumur – key facts

- **AOC granted**: 1962
- **Hectares in production**: 8
- **Production volume**: 190 hectolitres/25,000 bottles
- **Styles**: 100 per cent sweet white
- **Grapes**: Chenin Blanc
- **Soils**: tuffeau

The area covered by the Coteaux de Saumur appellation is, in fact, very large, comprising 28 communes of tuffeau-based soils around Saumur, but the average annual declaration area of wines under the AOC is less than 10 hectares, because most of the grapes will be made into dry wine. The conditions are less obviously well suited to making sweet wine compared to those of the Coteaux du Layon or Coteaux de l'Aubance. Furthermore, production rules are stricter, since no chaptalization is allowed. These are rare wines that have great ageing potential. A 22-year-old half-bottle tasted recently from Philippe Gourdon's Château la Tour Grise (a now-retired producer based in Le Puy Notre Dame) was ageing gracefully, the honeyed notes turning to barley sugar while losing none of its zesty freshness. It was lovely, and could doubtless last another decade or more.

Naturally, there are significant variations in the amount of sweet wine produced every year depending partly on the prevailing weather at harvest time and partly on the growers' choices about which wine styles they wish to make. Increasingly, the canniest growers are making the most of their options to produce dry wines as well as sweet.

Anjou sweet wine appellation production rules

Appellation	Permitted yield (hl/ha)	Chaptalization permitted?	Minimum residual sweetness (g/l)
Quarts de Chaume Grand Cru	20	No	85
Coteaux du Layon 1er Cru Chaume	25	No	80
Coteaux du Layon + village	30	Yes	51
Coteaux du Layon	35	Yes	34
Bonnezeaux	25	Yes	51
Anjou-Coteaux de la Loire	35	Yes	34
Coteaux de l'Aubance	35	Yes	34
Coteaux de Saumur	35	No	34

Notable Anjou producers

Château de Plaisance

Rochefort-sur-Loire

www.chateaudeplaisance.com

When Vanessa Cherruau decided to set up on her own, the plan she wrote for her dream winery read: '10 to 12 hectares, a great site, Anjou.' But when the 25-hectare Château de Plaisance came up for sale, she couldn't resist the challenge of dreaming on a bigger scale. 'I found an investor who wanted to help an environmental project without being involved,' she explains, establishing that she is the only one making the decisions here. A self-starter, Vanessa had no links in the wine business before enrolling herself as a student at the agriculture school in Angers. After work experience in Penedès, Germany, Austria and Champagne, she felt ready to go it alone.

Situated at the top of the hill of Chaume, her property comprises 20 of the 100 hectares of the sweet wine Chaume appellation, plus three hectares in Savennières and a little land elsewhere. 'The former owners were visionaries,' says Vanessa, 'converting to organic in 1995 and biodynamics in 2008.' The project has not been without its challenges. It took a year and a half to complete the sale and Vanessa found herself taking over the winery in September 2019, just in time for harvest and also just in time for the arrival of her first child. 'Luckily,' she says, 'my daughter was five days late.' She also had to galvanize a team of workers, including her husband and an old school friend, as most of the former employees had retired.

Since the challenge of the first vintage, the work has continued apace, including replanting neglected areas of the vineyard and dealing with one of the toughest vintages on record in 2021. But the domaine has rapidly been recognized for the quality of its wines. Vanessa is very clear on the style of wine she is trying to make: 'I like *al dente* Chenin: ripe and crunchy,' she says. This she achieves by hand harvesting everything and separating in the vineyard all the botrytized berries she doesn't want in her dry wines. 'Botrytis blurs the aromas,' she says. In all, there are five dry white cuvées, two reds and only one sweet wine from this sweet wine appellation. 'I'm not against sweet wine, but it needs to be special,' she says.

The different wines represent the varied soils of the domaine. Some are aged in oak of various sizes and ages, some in sandstone jars – there's

Vanessa Cherruau, a strong advocate for her appellation

no fixed recipe. The first of the range, L'Anjou Blanc, comes from a blend of soils and includes fruit from the younger vines of the domaine. Free from oak and bottled early, this is easy, crunchy, savoury, dry Chenin for immediate enjoyment. The largest cuvée, Ronceray, is a complex and mineral wine, with toasted hazelnut notes and a salty finish. La Grande Pièce comes from a single vineyard whose sunny exposure is amplified by the presence of large, rounded puddingstones. This is a richer style which nevertheless retains the vertical mid-palate, mineral intensity and flinty texture of the Anjou noir soils. In the opposite style is Zerzilles. These grapes would qualify for the Quarts de Chaume Grand Cru appellation if vinified as sweet wine. The steep, east-facing slope has a mixture of soils. Vanessa explains: 'Schist always gives power, spilite gives an upright volcanic quality and puddingstones roundness.' Zerzilles is a very complete wine, crisp and finely textured. Austere in its youth, with its rapier-like, mineral acidity, it's enough to make a Chenin lover swoon. The reds, while highly drinkable, are currently less convincing than the whites and, one feels, will improve over time. For now, dry whites are very much the focus: 'I don't like to pigeonhole things,' says Vanessa, 'but schist and Chenin Blanc is a great combination.'

In a short space of time, Vanessa has achieved a great deal, not least critical acclaim for her range of precise and expressive wines. As president of the Chaume appellation Vanessa, along with the other producers here and in Quarts de Chaume, is working towards the new Ronceray *cru* for dry wines made from these sweet appellations. 'I'd like to be able to put "Chaume Sec" on my bottles to show it's the same terroir,' she says, 'but I don't have the strength to fight the INAO. I prefer to spend my energy on planting hedges and increasing biodiversity in the vineyard.' Nonetheless, she remains enthusiastic about the future: 'Dry Anjou is the start of history,' she says, 'we can invent things, try and fail – it's very free and very exciting.'

Château de Suronde

Rochefort-sur-Loire

www.chateaudesuronde.com

This 7-hectare estate in a bucolic spot on the hill of Chaume has been owned since 2016 by former management consultant Kathleen Van den Berghe and her husband Sigurd Mareels, who also own the much larger estate of Château de Minière in Bourgueil. Both operations are impressive in their scope and dynamism, particularly since the couple is still based in Belgium. Kathleen appears to have more hours in her day than the rest of us. When she is not troubleshooting in the winery or travelling between one or other of her estates, she is often seen pouring her wines at fairs for potential customers.

Suronde is more than just a wine estate. Billed as 'a place of creation', every year it welcomes a different artist to spend time living at the château to create art that Kathleen selects for the Suronde label of the vintage. At other times, the château and a renovated neighbouring property are available for anyone to rent. The château enjoys a particularly lovely outlook across the vines and some of the 10 hectares of unplanted land and forest surrounding the property.

Château de Suronde is almost entirely situated in the Quarts de Chaume Grand Cru, with just a small parcel in the Coteaux du Layon Premier Cru Chaume. Nevertheless, as is the fashion, most of the wine produced here (up to 75 per cent) is dry, although Kathleen says she is happy to see a growing interest in the sweet wines, especially Coteaux du Layon. Any dry wine is currently classified as Anjou blanc now, but will most likely be Ronceray Premier Cru in the future. Kathleen, who expresses a preference for 'skinny wines', says she is aiming for a style of wine that is 'lean and clean, linear and pure' for her entry-level L'Esquisse. Made from the fruit of younger vines of the estate, it is fermented in stainless steel and bottled the following spring. It shows a clear, pure acidity and upright Chenin fruit. The second dry wine, La Création, comes from 50- to 60-year-old vines, and the top dry wine, L'Oeuvre, comes from vines at the heart of the Quarts de Chaume vineyard. Early vintages of these wines were fermented in stainless steel, but from 2020 Kathleen has switched to large oak for La Création and concrete egg for L'Oeuvre. This change has led to more approachability in the wines, and more of a sensation of roundness (especially for L'Oeuvre), although they remain true to the house style, which is light and elegant, with very crisp acidity. The relatively small amount of sweet wine shows beautiful fruit clarity. Château de Suronde Coteaux du Layon is a pure, fresh and very drinkable wine, with between 50 and 70 grams per litre of residual sugar, while the Quarts de Chaume shows the natural brilliance this site can achieve when the grapes are given the right attention. It has wonderful balance and a dreamy texture. Suronde is a quality estate where the wines are getting better every year.

Château Pierre-Bise

Beaulieu-sur-Layon

www.chateaupierrebise.fr

This 45-hectare domaine in the heart of the Layon vineyards began as a much smaller affair in 1959, when Pierre Papin bought the château with 9 hectares of vines. It was his son, Claude, who expanded the domaine to its current size when he took over in the 1970s, including the acquisition of his wife's family vineyards in the 1980s. Today the array of plots covers an area that extends across 14 kilometres and includes many of the great terroirs of Anjou, including Savennières, Chaume and Quarts de Chaume.

Claude Papin was ahead of his time in adopting natural methods in the vineyard and heavily influenced by the work of Japanese farmer and philosopher, Fukuoka Masanobu (1913–2008), who promoted a no-till, observational approach to farming. Claude was already planting cover crops in the 1970s and the domaine finally became certified organic in 2019.

Today, Claude's son René manages the domaine, producing a wide range of quality wines across the spectrum of Anjou appellations, including a more than creditable Crémant de Loire. Chenin Blanc accounts for more than half of the production and, unsurprisingly, most of the renown of the domaine. The sweet wines are particularly delicious, showing the balance and grace of Chenin grown and harvested with care on the complex soils of Anjou. The Coteaux du Layon wines are masterful examples of what this appellation can achieve.

Domaine Belargus

Saint-Lambert-du-Lattay

www.belargus.com

Domaine Belargus is a recent venture from Ivan Massonnat, a new vigneron who has done much to change perceptions about Chenin Blanc from Anjou since he created his domaine in 2018. In a bold move, he released the first vintage of his top cuvée, Treilles, at 100 euros a bottle – and sold the lot. 'My aim was to break the glass ceiling from day one,' he says, referring to Anjou's lacklustre image as a source for high-quality wine. It looks like he's managed it. With 24 hectares of vines, clever marketing and crack team of eager young talent, Domaine Belargus has quickly become a reference for top Chenin Blanc, grown on some of the best terroirs in Anjou.

Ivan has made the transition from finance to fine wine look easy – but this is a project that was a long time in the planning. He spent 20 years learning about wine in his spare time and visiting vineyards from his Paris base. He says he never had any doubt that the Loire would be the location for his next challenge. He had owned a house in nearby Chinon for 20 years, so knew the area well. He chose Anjou, he says, because 'it is the home of the oldest, smallest quality appellations for Chenin – and the region was clearly underrated and misunderstood.' This view was confirmed when he started looking for land to buy, in 2016. 'There were a lot of estates for sale,' he says, 'I couldn't believe it.'

After two years of searching, he was introduced to veteran Anjou winemaker, Jo Pithon, who was retiring and looking to sell his vineyards. Visiting the 7-hectare Coteau des Treilles vineyard in February 2018, Ivan's mind was made up. He agreed to buy it on the spot – a *coup de coeur*, as the French say (see box, p. 94). 'I decided there and then,' he says (remembering not only the date, but the time of his epiphany), 'my project starts here.'

The reclaiming of the Coteau des Treilles

You could be forgiven for thinking you've taken a wrong turn if you visit the Coteau des Treilles vineyard. Down a narrow country lane, darkened in summer by dense trees reflecting their cool, deep green into the peaceful Layon below, the vineyard appears out of nowhere. It's an arresting sight: a hill of vines bathed in sunlight looming up from the road and topped by an outcrop of rock.

Old postcard of Coteau des Treilles

The vineyard was developed in recent times by now-retired local winemaker Jo Pithon, who owned vines just across the river (Les Bonnes Blanches, in the upper right of the photo). He says he had often wondered why this steep, south-facing site was not a vineyard. His curiosity was further aroused when he found the postcard above from the 1920s showing vines growing there. It appears the vineyard was abandoned after the twin ravages of phylloxera and the Second World War. One has to assume that the local farmers decided there were easier places to work – and perhaps that life was just too short to bother with this place. But Jo had other ideas. Inspired by visiting the vertiginous vineyards of Egon Müller in the Mosel, he invited the legendary German winemaker to come and give his opinion on the potential of this overgrown slope. The great man's advice was unequivocal: 'Plant it!' This was easier said than done. Jo set about finding the dozens of owners of the overgrown site, bought the land for a song and found some existing loyal customers who were willing to help him realise the new project. After clearing the impenetrable tangle of trees and brambles he managed to replant the site in 2008.

The revived Coteau des Treilles vineyard

The climate in this vineyard is on average 2°C warmer than in the surrounding villages. The complex geological profile includes four main types of rock: schist, spilite (a hard volcanic rock), grey limestone and conglomerate puddingstones. New owner Ivan Massonnat describes it as a 'micro chaos' of soil types. The combination of this geological complexity, the steepness of the slope and its south-facing exposure creates an unusual microcosm that has long attracted botanists – it is also a classified bird reserve. The soil is in parts acidic and in parts alkaline, so an unusual profusion of insects and flowers flourishes here: 773 species of flower and 1336 animal species have been identified. From one of these comes the name of the new domaine. The little *Lysandra bellargus* butterfly, also known as Adonis blue, is easily spotted by its cornflower-blue wings. Its presence is proof of limestone soils, which is unusual in this part of Anjou.

Over the next six months, Ivan made further purchases to add to the domaine's portfolio, in the Savennières and Quarts de Chaume appellations. An indication of the lack of interest in Anjou is that the 10 hectares of Quarts de Chaume Grand Cru vineyard that he bought (which amounts to one-quarter of the appellation) had been on the market for a decade. Ivan says, 'People were looking at the price and saying it was ten times more expensive than the vineyard touching it; I was looking at it from an absolute standpoint: it was cheap compared to anything else of similar quality, from Burgundy or Alsace, for example.'

From the point of buying the vineyards, everything happened very quickly, Ivan recalls. 'I signed the papers and five days later we were picking the grapes. And we released 14 cuvées the first year.' The wines, all exclusively from Chenin Blanc, are mostly dry. Other than the two Savennières wines, the rest are classified as Anjou blanc, since the vines grow on land classified in the sweet wine AOCs of Quarts de Chaume and Coteaux du Layon, and these appellations can only be claimed for sweet wines. Most of the cuvées are of individual vineyard sites, reflecting the terroir where they are grown. All wines are made in the same way, with slow, whole-bunch pressing followed by hands-off vinification and ageing in large oak vats, with no *bâtonnage*. The main difference between the handling of the wines is the length of ageing, which can be between one and three years.

Ivan Massonnat with some of his young vines

The wines are delicious, and offer a fascinating gustatory tour of the soils of Anjou noir. Particular favourites include Ruchères, a Savennières of silken power that grows on a tiny plot of purple schist below the Roche-aux-Moines vineyard, and Quarts, at the heart of the Quarts de Chaume appellation, which produces sublimely elegant wines of minerality and freshness, whether dry or sweet. Also of note is the icon, Treilles, a wine of tension and power that amply shows the complexity of its origins. Across the Layon from Treilles is the north-facing Bonnes Blanches, a late-ripening site that produces a quietly elegant wine. Adjoining either side of the Quarts vineyard are two further plots in the Quarts de Chaume appellation, vinified separately. Rouères comes from the warmest part of the slope, a south-east-facing site that produces richer wines; Veau, to the west, is cooler, with thin soils that produce a more linear, mineral expression of site.

Having achieved a lot in just a few years, one might expect Ivan Massonnat to be taking a breather. Far from it. He is planting new sites on more Anjou terroirs nearby, and has meanwhile bought another estate, in Chinon. The first vintage of his Domaine de Beauséjour has yet to see the light of day. With this new project, he says, he has learned from his experience with Belargus that he can afford to take his time. Clearly this is a new life that suits him, and Ivan is filled with enthusiasm for the future, stressing, 'This is not an ego trip or a retirement project. It's a new professional challenge. I'm ahead of my business plan – but we are nowhere near our destination.'

Domaine de Bablut

Brissac

www.bablut.fr

The Daviau family has been farming the land around the town of Brissac since 1545; for much of that time grapes were just one of their crops, their main activity being cereals. They originally owned five mills on the slopes overlooking the Aubance. Towards the end of the nineteenth century they decided to specialize in viticulture and today the domaine is in the hands of Antoine Daviau and his son Christophe, who joined his father in 2018, changing track from a career in law to retrain in order to take over from his father when he retires. Both father and son have experience working outside the region, Antoine in Australia and Christophe in Sauternes, Châteauneuf-du-Pape and New Zealand.

Antoine was an early adopter of organic viticulture and had trialled biodynamic methods since the early 1990s. The estate was certified biodynamic in 2000. Preservation of the environment and building biodiversity are important priorities for the family; they are planting trees in and around the vines and grass between the rows, with a flock of sheep to graze there.

At 55 hectares, this is a large domaine and a wide range of wines is produced, including the appellations Anjou, Anjou Brissac, Coteaux de l'Aubance, Rosé de Loire, Cabernet d'Anjou and Crémant de Loire, as well as some *vins de soif* bottled as IGP Val de Loire. Of particular interest are the reds from the small appellation of Anjou Brissac, an area of slopes from 50 to 90 metres' altitude composed mainly of schists, with quartz, sandstone, some volcanic rocks, sand and silt, but also some parcels of limestone, as this terroir is on the border between the Anjou noir and Anjou blanc soils. Yields are lower than for Anjou Villages and the wines, always red, have greater concentration. The domaine makes three wines in this appellation, showing the interesting effect of these varying soils. Petra Alba comes from Cabernet Franc grown on limestone, a full-bodied wine with restrained, powdery tannins. Rocca Nigra is an unusual wine, a schist-grown Cabernet Sauvignon. Christophe says, 'The grapes need drainage but not stress.' They are picked when very ripe. In good vintages, this gives a firm wine with good depth of mouth-filling fruit and flavours of liquorice, tobacco and smoke. It has the mineral and slightly saline notes common to the whites grown on this type of terroir. A fascinating, powerful wine very much of its place, it will last a decade or more. The Daviaus also make a delicious Crémant de Loire from Chardonnay, Grolleau Noir, Chenin and Cabernet Franc that ages for three years on its lees. The grapes grow on a mixture of Anjou noir and Anjou blanc soils, which gives added complexity to the blend.

Domaine des Baumard

Rochefort-sur-Loire

www.baumard.fr

Domaine des Baumard can trace generations of family vine growing going back to 1634, but this is far from being a traditional estate. Jean Baumard, who started the business in its current form in 1955, had no hesitation in applying radical new ideas to growing his grapes and making his wines.

A scientist by training, Jean studied oenology in Dijon and Bordeaux and in time became a professor of viticulture and oenology himself at the local university in Angers. As head of the family domaine, he applied himself to experimental techniques of grape growing, allowing the vines more space to grow and breathe. His son, Florent, who runs the estate today, refers to this method of grape growing as VHL – Vignes Hautes et Larges – and if you walk through the vineyards of Savennières there is no mistaking the Baumard vines, which stand out from those around them at two metres tall and with wide, three-metre spacing between the rows. Appellation laws have strict rules about vine spacing and density (and indeed most other aspects of grape growing and winemaking), but there are always exceptions made for experimental vines, and these vines have somehow been allowed to stay. Florent Baumard explains that the idea is to avoid stressing the vines, but instead to give them the ideal conditions in which to thrive.

It seems as if everything was considered from a fresh perspective from the start at Domaine des Baumard, with little attention paid to convention unless it suited Jean's ideas. He was an early adopter of the now-commonplace practice of using non-botrytized grapes to make Savennières, preferring the more mineral style to express the purity of the fruit. This is not always easy to do, however. 'The difficulty with Chenin Blanc is that it doesn't ripen evenly,' says Florent, 'On the same plant you can have five or six different stages of ripening and each gives a different taste. That's something you can't learn in a book.' The idea, he says, is to get perfect ripeness.

The domaine was also the first to own land on both sides of the Loire, in Quarts de Chaume and Savennières. When it comes to winemaking, no oak is used, and everything is fermented in stainless steel. 'I have used wood in the past,' says Florent, 'even though my father told me not to. But when you have a great terroir, it's a shame to mask it with oak.' As with most domaines these days, malolactic fermentation is also common. 'I'm neither for it nor against it,' Florent says. 'We are looking to make wines that are good to drink and that will age a long time.' He says he has noticed that now the malolactic fermentation takes place earlier, 'during or even before the alcoholic fermentation,' which has the advantage of producing wines with less obvious malolactic character.

Perhaps the most radical decision of all was to bottle under screw cap. The first vintage sealed using this closure was 2003. 'I wish I'd done it sooner,' says Florent. He is the only Savennières producer and one of very few in the Loire to bottle fine wine this way. His rationale? 'We are trying to make wines that are fine, delicate and precise.'

The two Savennières cuvées of the domaine are Clos du Papillon, from the vineyard of the same name, and Clos de Saint Yves. The two plots are close to each other. Both produce harmonious and approachable wines with pristine fruit. The Clos du Papillon is slightly more expensive ('just because there's less of it', says Florent). It is slightly richer and broader in style than the mineral and slightly more vertical Clos de Saint Yves that grows on purple schist. While the Savennières wines are the flagships here, the domaine also produces wines in the Quarts de Chaume and Anjou blanc appellations.

Domaine Ogereau

Saint-Lambert-du-Lattay
www.domaineogereau.com

Emmanuel Ogereau is the fifth generation in his family to farm the terroirs of Anjou in the valley of the Layon; his great-grandfather was the first in the family to plant vines after the phylloxera crisis at the turn of the twentieth century. Known for many years as a quality address for sweet wines from the great vineyards of Chaume and Quarts de Chaume, today the domaine's focus is on dry whites, which now account for 70 per cent of production – a 180-degree turnaround. 'It's not easy to do,' says Emmanuel, recalling that it took six attempts (that's six vintages) to make a dry wine they were happy with from their Quarts de Chaume vineyard, La Martinière. For the fortunate drinker of Ogereau wines, though, there is nothing in this range that hints at difficulty or struggle: there is an easy-going and effortless quality in these wines and it is tempting to think that this has something to do with the easy-going winemaker with the infectious smile at the helm.

Sixteen different cuvées are made, across all colours and styles. The wines are consistently good, and all are highly drinkable, but the standouts are unquestionably the Chenins. These are very special wines that draw into focus the unique and geologically diverse terroirs of the Layon Valley and Savennières. The work in the vineyard is organic and biodynamic and everything is harvested by hand. After gentle pressing, the juice settles before going to tank or wood – or occasionally, amphorae – for fermentation and ageing.

The largest cuvée of the domaine, 'En Chenin' is a tank-fermented, early-bottled wine that encapsulates all that is good about Chenin from Anjou noir, with its fresh, mineral intensity and bitter finish.

Domaine Ogereau vines growing on volcanic spilite

Bonnes Blanches is a flagship vineyard on the left bank of the Layon originally planted by Emmanuel's father for sweet wines. Since 2017, the domaine has made both sweet and dry wines from this site. The vines grow on a layer of clay with a relatively soft and friable schist rock beneath. The clay in the soil gives a broad base to the wine, but it retains freshness and precision. For this reason, Emmanuel describes the dry Bonnes Blanches as 'a triangular wine'. Vent de Spilite is another dry Chenin of an entirely different dimension. In this case, the grapes grow in the Pierre Bise vineyard, a high and windy volcanic plateau that sits on the Layon geological fault overlooking the Bonnes Blanches across the river. A firm and austere expression of Chenin derives from this dry site where the soil is rich in iron and magnesium. The name Vent de Spilite comes from the unusual vein of spilite found here, clearly visible on the surface of the vineyard (there is also a quarry for extracting this rock next to the vines). It is a mineral and tightly focused wine, with juicy bitterness and saline freshness. Apart from

Emmanuel Ogereau with his mouthwatering Vent de Spilite

a few rows of older vines, the vineyard was planted mostly in 2013, so the quality of wine from this extraordinary site will only improve.

Three Savennières cuvées are relatively recent additions to the Ogereau portfolio. The young vine La Saponaire wine is a pure, fine, delicate and elegant expression of Chenin. The Clos Le Grand Beaupréau has more fruit concentration but is equally delicate, with a soft, silky texture and excellent balance. Within this same vineyard, Emmanuel identified that vines growing on a vein of sandstone gave grapes of a quite different character. The result was a new cuvée (from 2014), L'Enthousiasme, a more muscular wine with rich fruit, gorgeous intensity of flavour and a lightly tannic texture.

The sweet wines are no less impressive, combining easy drinkability with botrytis complexity and silky, honeyed fruit. They manage to be light and pure, rich and complex, all at the same time. More seemingly effortless stuff.

Ogereau wines are characterized by their precision, texture and length, rather than overt fruit expression. Emmanuel says, 'Aromatically, Chenin is very shy here. I'm focused on mouthfeel, texture, balance and a profile that reflects the terroir.' It's a winning formula.

Domaine Patrick Baudouin

Princé, Chaudefonds-sur-Layon

www.patrick-baudouin.com

'I'm always getting myself into fights I've lost before I even start,' Patrick Baudouin tells me, with a wry smile. The truth is, he likes a fight. And despite his protestations to the contrary, some of them he even manages to win.

Patrick Baudouin's life was on quite a different path before he returned to his family's vineyards in 1990. Although his family origins are in Anjou, Patrick lived in Paris until he was 40, immersed in a life of books. But then he discovered wine and came home to pursue his new passion. By this time, though, most of the 13 hectares of vineyards acquired by his great-grandparents at the beginning of the twentieth century had been rented out. Starting with the remaining 1.5 hectares, he gradually built the domaine back up, developing fallow land and reclaiming and renting vineyards where possible. Today he farms 10 hectares of Chenin Blanc and 3.5 hectares of Cabernets Sauvignon and Franc, plus a tiny amount of Grolleau Noir, across several sites. All of the land is on the Anjou noir, schist-based terroir, the old soils of the Armorican Massif.

Patrick's vine holdings cover the appellations Anjou, Savennières, Coteaux du Layon and Quarts de Chaume. Like many other producers, he has been focusing his efforts on dry wines since 2000, and is actively trying to promote the top sites in Anjou as *crus* for dry white wines.

He has been equally active in defending the quality of sweet wines, however, where his lobbying has been orientated towards improving quality by trying to instigate a ban on chaptalization. This is still permitted for Coteaux du Layon wines if potential alcohol levels are low at harvest time. Another battle he is happy to recall is for an individual, extremely sweet, wine he made that was refused the Quarts de Chaume appellation because the alcohol level of 9% abv was below the prescribed minimum of 11%, even though the wine, lusciously sweet, was of high quality. Eventually, he prevailed on that one: unlike the other sweet wine appellations of Anjou, there is now no minimum alcohol level for the Quarts de Chaume Grand Cru.

Patrick's wines are organic, but he rejects biodynamic principles as 'a distraction'. He is always happy to take the time to explain his point of view on this and the many other areas on which he holds a strong opinion, always with a smile on his face. When it comes to so-called natural wines, he points out, 'deciding not to do something is a choice in itself'. Nevertheless, he uses predominantly natural methods in his vineyards and winery.

When I visit, a horse-drawn plough is working the soils, sheep graze in the vineyard, and the chickens climbing the steps into the house Patrick addresses by name. The winery was purpose-built using natural materials and incorporating a Canadian well (a type of underground heat exchanger) for temperature control. It's no surprise to learn that the musts are never chaptalized here, and sulphur usage is kept to a minimum. Fermentations can be long and continue through to the spring following harvest because Patrick wants to limit energy usage in the winery. He

Inveterate campaigner Patrick Baudouin

always aims for the whites to go through malolactic fermentation which, he says, 'adds to the depth of the wines and helps them to age better'.

Patrick is aiming to express the individual terroirs in his wines, achieved by low yields and minimal intervention in the winery. For me, the sweet wines are the highlights of the range, but the dry whites from old vines, such as Le Cornillard, show great depth of flavour and tingling minerality. Although, by his own admission, these are not red wine terroirs, the domaine's reds are better than most. La Grolle (made from Grolleau) is perfumed and fresh, while the Cabernet Franc La Fresnaye has satisfyingly juicy, crunchy fruit. The secret, says Patrick, is 'low yields and ripeness to avoid the pyrazines'. Like the man himself, this is a domaine whose wines are never boring.

Grange Saint-Sauveur

Blaison Gohier

This unmarked winery is tucked away in one of the remoter corners of Anjou – which is probably just how Alice and Antoine Pouponneau like it. The friendly but media-shy couple have no website or listed phone number, and the labels of their wines – all Vins de France – carry few clues as to their origins. Antoine points out that the flat-bottomed boat drawn on their labels is traditional to the Loire (although he does admit, you need to know this – '*il faut savoir*'), and the name of the tiny village where the vineyards are located, Le Thoureil, is also noted (but again, '*il faut savoir*'), as is the number of the department (but, yes, '*il faut savoir*'). And the back label gives no information at all because it is not there. 'I can't stand back labels,' says Alice. 'I'm not going to start telling people how to drink our wines.'

But their names are there in black and white, and perhaps you can afford to be a little anonymous with your labels when you are as in demand as Antoine Pouponneau. An oenologist with training from Dijon and decades of winemaking experience under his belt at Clos Canarelli in Corsica and La Tour du Bon in Bandol, he has more recently become sought after as a consultant for the likes of Cheval Blanc, Châteaux Latour and Pavie in Bordeaux, and Dalla Valle in Napa. His particular expertise is in making wines that combine precision with a minimum level of additives, in particular industrial yeasts and sulphur. This is knowledge he now has the opportunity to refine back in his Anjou home, in the couple's garagiste-style winery. And despite his apparent anonymity, his customers track him down easily enough.

Alice and Antoine established Grange Saint-Sauveur from their home here in 2017. Saint-Sauveur is the name of the tiny fifth-century chapel sandwiched between their traditional stone home and the barn ('*grange*') where the winemaking takes place. Everything is at a comfortably human scale and at the time of my visit, in October, the wines fermenting in barrel were also looking comfy, cosily tucked up under a selection of blankets and bedcovers from the house. There's nothing flashy about this place, and certainly nothing as fancy as temperature-controlled vats. The wines ferment in barrels or small stainless steel vats or a combination of both, with native yeasts and no sulphur added during the fermentations. Finished wines undergo long ageing in the barrel-lined chapel (could there be a better place for wine to rest?) and, often, further ageing in bottle. There is no fining or filtration here, but Antoine is not a natural wine zealot and will add an amount of sulphur dioxide at bottling as needed, although for wines he feels are especially well balanced, this might be zero.

The couple's 6.5 hectares of vines, all growing above the picturesque riverside village of Le Thoureil, include a mixture of local varieties such as Grolleaus Noir and Gris, Gamay and Pineau d'Aunis, as well as Chenin and Cabernet Franc. Sometimes varieties grow together in the vineyard and almost all were propagated by massal selection long ago. Yields are uniformly low from vines as old as these. The wines include a fine, mineral, structured and mouthwatering Chenin Blanc – Centenaire – from 100-year-old vines, and a lusciously fruity yet uprightly structured Cabernet Franc – Huguette – from 70-year-old vines. The other wines are blends, which are their real passion: 'It's a difficult job and it would be sad not to enjoy it – it's very gratifying to construct a wine,' says Antoine. Les Arceaux, a blend of Grolleau Noir and Grolleau Gris, is a remarkable rosé unlike any other, aromatic, structured and bone dry, with savoury delicacy and extraordinary length. Clandestine, a blend of Grolleau Noir and Cabernet Franc, is a wine of enveloping aroma and arresting purity, a gentle blend of dark chocolate and cherries, with fabulous balance and a long, spicy finish. The short maceration for Le Martray brings the fruit to the fore. It is a dashing and youthful blend of Grolleaus Noir and Gris and Gamay, with a succulent red-fruit character.

The wines of Grange Saint-Sauveur carry a signature of lightness, balance and freshness. Erring towards understatement, but always elegant and invariably interesting, they are some of the best expressions of the *ligérien* style, from vines that have been looking out over the Loire since long before many of their drinkers were born.

Moulin Touchais

Doué-la-Fontaine

www.touchais.fr

Moulin Touchais is one of the success stories of the post-war era in Anjou. The Touchais business was founded in 1787, but it was Joseph Touchais who developed this singular wine following the privations of the Second World War. Moulin Touchais, a wine from the Coteaux du Layon, had been made since the 1930s. According to export manager Frederik Wilbrenninck, the story of the wine as we know it today is that during the war Joseph Touchais had the foresight to wall up his precious inventory of sweet wine to save it from being drunk by the invading army. After the war, prices for such luxuries had collapsed, so he waited for the market to pick up before he released these wines back onto the market. He was able to do this because he had meanwhile set his mind to another Anjou project, a branded rosé called Cuisse de Bergère, which translates as 'Thigh of the Shepherdess'. This wine, appropriately (or perhaps inappropriately) labelled with a *bergère* showing a good deal of *cuisse*, was an instant hit, selling as many as 10 million bottles a year in the 1950s. Joseph was therefore able to take his time to wait for the economy to pick up before going back to his Moulin Touchais bottles. By the time he did, ten years had passed and he noticed that the aged wine was greatly improved. Since then, Moulin Touchais has never been released until it is at least a decade old.

There can't be many domaines where you can buy a large selection of vintages going back as far as the 1950s. I recently tasted 45 vintages of Moulin Touchais, plus La Réserve du Centenaire, a non-vintage cuvée of wines from 1849 to 1899 that were blended and rebottled in the 1950s by Joseph Touchais. It is a revelation to see how well these medium-sweet wines can evolve over time, and to feel the effects of the vintage decades on. Moulin Touchais offers a rare taste of history.

Terre de l'Elu

Saint-Aubin-de-Luigné

www.terredelelu.com

For a taste of the more creative side of winemaking in Anjou, Terre de l'Elu is a good place to start. Here, Brittany natives Thomas and

Charlotte Carsin make a range of interesting, naturally focused wines, often from blends of multiple grape varieties.

Thomas is an agronomist who previously worked as a viticultural consultant in California, Provence and Champagne, specializing in soil analysis. In 2008, the couple decided to take the plunge and buy their own wine estate, which brought them to Anjou. The 22-hectare Terre de l'Elu was established in the 1950s and its vines are situated in the warm and dry Layon Valley. All of the vines are on the right bank of the river, which benefits from southern exposure. As one might expect, much attention is given to the vineyard, which is run organically, and all of the fruit is hand-picked.

Because the Carsins use some varieties outside of the appellation and blend some of their wines, which can also taste atypical (another problem if one wishes to make wine within the appellation system) they decided to dispense with it altogether, so all of the wines are simply classified as Vin de France. Production is split evenly between red and white wines. Fermentations are on the natural side, often with very little use of sulphur. Roc'h Avel is a bright and well-balanced 50/50 blend of Sauvignon Blanc and Chenin, co-fermented and aged for a time on lees. The Sauvignon gives it a spiky edge that is softened by the breadth of Chenin. Bastingage is a rich, 100 per cent Chenin Blanc wine, with depth and substance. Ephata is another Chenin wine, this time aged in amphorae. It has intense minerality but is also open, soft and rich, with a long-honeyed finish – an interesting, different kind of wine. Of the reds, Maupiti is a very gluggable Gamay–Cabernet Franc blend, scented and fresh with a bit of bite on the finish. At the opposite end of the spectrum, the domaine's top red cuvée, Magellan, is a firm and concentrated wine that comes from 80-year-old Cabernet Franc vines. The fruit is fermented in amphorae, then pressed and aged for two years in barrels. While Thomas is the main winemaker, Esperance, a Pineau d'Aunis, is made by Charlotte. It is a ripe, elegant, gently spicy wine, quite unlike the more typically peppery examples found elsewhere.

Charlotte and Thomas are in the process of planting trees to increase biodiversity and their latest project, Entre Deux Mondes, is a wine Charlotte says is to 'represent the Anjou red of the future' and to allow their customers to 'drink a place', the place being a 1-hectare clos where the component parts of Cabernet Franc, Grolleau Gris, Pineau d'Aunis, Chenin Blanc and Gamay are all planted together.

Other producers to try

Château de Bois Brinçon, Château de Bonnezeaux, Château de Passavant, Château du Breil, Clau de Nell (for the Burgundian-style whites), Domäne Vincendeau, Domaine Cady, Domaine de Bois Mozé, Domaine de la Bergerie, Domaine des Forges, Domaine des Hardières, Domaine des Petits Quarts (Bonnezeaux), Domaine FL (see Savennières), Domaine Juchépie, Donatien Bahuaud, Eric Morgat (see Savennières), Le Clos Galerne, Terra Vita Vinum, Thibaud Boudignon (see Savennières).

When in Anjou – places to eat

La Route du Sel in Le Thoureil

The city of Angers has several good options for wine lovers, including the relaxed Une Fille & Des Quilles (unefilleetdesquilles.com). In the countryside there is less choice, but there are two excellent restaurants. The first, La Table de la Bergerie (www.latable-bergerie.fr), is in the heart of the Layon vineyards, and is owned by the Domaine de la Bergerie winery. The Michelin-starred restaurant features a simple, calm and airy space with a terrace opening on to the vineyard. There is a limited selection of dishes, but each is prepared to perfection. The extensive wine list includes many of the wines of Domaine de la Bergerie available by the glass. There is also the possibility to stay here at the gîte, which sleeps eight people. In the pretty riverside village of Le Thoureil is the delightful restaurant La Route du Sel (www.authoureil.fr). The tree-shaded garden to the front, set with tables, benches and parasols, is a lovely spot to while away a few hours on a summer's afternoon. The wine list is extensive and reasonably priced and the friendly staff welcome guests for lunch, dinner, or simply a drink in the garden, where they serve particularly good snacks to accompany a local glass. The menu features fish from the Loire and meat from the surrounding area, home-made pasta and fresh salads. In the cooler months the cosy dining room is a popular place and usually full. A mournful basset hound adds to the laid-back ambience.

SAVENNIÈRES

The vineyards of Savennières lie on the north bank of the Loire behind the tiny village of the same name. The Savennières appellation extends to the east and west to include the communes of Bouchemaine and La Possonière respectively. From the south the village is reached by an improbably narrow road bridge that spans a wide stretch of the river, passing through the charming island of Béhuard en route. North of the river the climate is always a little cooler than to the south, where most of Anjou's vineyards lie, but in Savennières the vines enjoy a privileged position of mostly south-facing sites, where the climate is moderated by currents of air flowing up from the river below. These vineyards, which follow the line of the river for around 6 kilometres, have for many hundreds of years been recognized for the excellent quality of the wines they produce from Chenin Blanc.

The monks from religious orders in Angers, 15 kilometres distant, were the first to be known to plant vines here. Records show that the monks of the Abbey of Saint Nicolas planted the Roche aux Moines vineyard in 1140, and the Coulée de Serrant can claim a similarly ancient lineage of wine growing, witnessed by its own Cistercian monastery next to the vineyard.

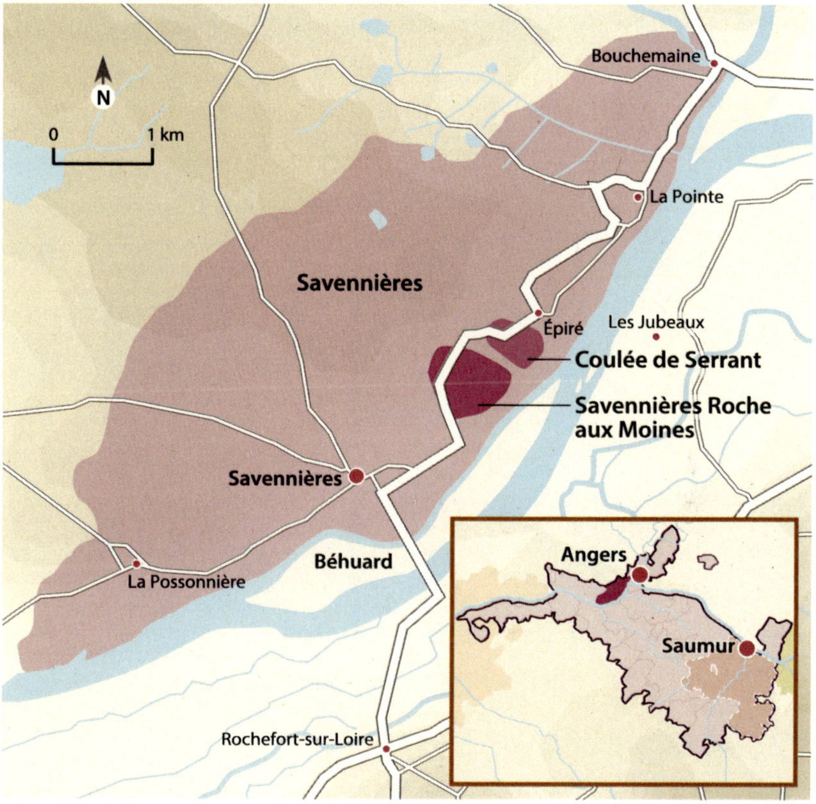

Map 7: Savennières

Further wine-growing developments came courtesy of the Angers bourgeoisie, who built themselves country homes there, a trend that continued with the construction by Parisian financiers of 17 châteaux, each with its own large park and vineyard, in the 1800s. Evelyne de Pontbriand, president of the Savennières appellation, explains the special way the vineyards were constructed. 'The appellation was designed holistically: the vineyard is always linked by a path, with woods, shrubs, lawns and water. This is unusual in the idea of a wine château. All the houses were closely linked to the farms. And this has remained almost untouched.'

In 1952 the Savennières appellation was officially recognized, defining the growing area at 350 hectares. However, the challenging nature of the terrain means it has been difficult to develop the vineyard to the full, a situation compounded by construction of habitations. Today less than half of the potential vineyard area is planted, although some

Vines growing on purple schist, Savennières

newcomers, such as Eric Morgat and Thibaud Boudignon, have revived interest in planting some of these abandoned areas.

Unlike the appellations across the river in Anjou, the vignerons of Savennières have never lost the option to make both sweet and dry wines. Although it had historically been known for its dry wines as well as sweeter ones, the original INAO regulations did not specify that wines needed to be dry or sweet. At the same time as the sweet appellations of the Layon were required to commit to making sweet wines only in 1996, the INAO requested that Savennières commit in a similar fashion to making dry styles only as, by then, this was the style for which the appellation was best known. But the Savennières producers were having none of it. Led by Evelyne de Pontbriand's mother, the late Michèle Bazin de Jessey (who was at the time owner of Domaine du Closel and president of the appellation), the growers said '*non*' to giving up the rights accorded to them on the creation of the appellation. So, although it is not so common to see sweet, *demi-sec* and even *moelleux* or *doux*, Savennières wines can still be made under the original appellation rules today.

Savennières – key facts

- **AOC granted**: 1952
- **Hectares in production**: 140
- **Production volume**: 4,300 hectolitres/570,000 bottles
- **Styles**: still white, mostly dry but may be *demi-sec* (8–18 grams per litre of residual sugar), *moelleux* (18–45 grams per litre), or *doux* (more than 45 grams per litre)
- **Grapes**: Chenin Blanc
- **Soils**: shallow schists, volcanic rock, veins of quartz, wind-blown sand

Savennières Roche aux Moines – key facts

- **AOC granted**: 2011
- **Hectares in production**: 18
- **Production volume**: 370 hectolitres/49,000 bottles
- **Styles**: still white, mostly dry
- **Grapes**: Chenin Blanc
- **Soils**: as Savennières, but tending to be stonier

Coulée de Serrant – key facts

- **AOC granted**: 2015
- **Hectares in production**: 7
- **Production volume**: 160 hectolitres/21,000 bottles
- **Styles**: still white, mostly dry
- **Grapes**: Chenin Blanc
- **Soils**: as Savennières, but stonier

While Nicolas Joly, owner of La Coulée de Serrant, is probably the French vigneron best known for his vigorous promotion of the virtues of biodynamic viticulture, 80 per cent of winegrowers in Savennières are, in any case, organic. Evelyne de Pontbriand says, 'When I took over from my mother, she told me I'd never need insecticide, but nobody knew why. Now we know it's because we have a very high population of bats.' These bats thrive in a succession of green corridors known as '*coulées*'. At night, they fly up and down these valleys eating the insects that can cause damage to the vines. Now that this is understood, the vignerons are planting more trees and bushes to encourage even more bats in the vineyards, something that is being copied in nearby appellations such as Chaume. 'The study of bats made everybody conscious of the value of the landscape,' says Evelyne.

The soils of Savennières are mostly schist, with wind-blown sand on the top. These are often very shallow, poor soils that are difficult to work. Eric Morgat, who makes wine in Savennières and Anjou, considers Chenins from Savennières to have more tannin, bitterness and austerity compared to those to the south of the river which are, in his view, 'more exotic and '*gourmand*'. He says, 'Savennières is difficult to understand, but interesting to work.'

The wine styles of Savennières have changed radically in recent years. As recently as 20 years ago grapes were harvested late, with winemakers waiting until the crop was affected by botrytis. At La Coulée de Serrant, Nicolas Joly and his daughter Virginie continue to work in this time-honoured fashion, but virtually all of the 30 or more other Savennières producers now avoid botrytis as far as possible. Picking earlier results in wines that are lighter in alcohol and approachable sooner. One thing that has not changed is that all the grapes in the three appellations are harvested by hand, a requirement of the regulations. Nowadays, most wines also go through malolactic fermentation, which further adds to their approachability.

Notable producers

Domaine aux Moines

La Roche aux Moines

www.domaine-aux-moines.com

The Domaine aux Moines is one of those properties that makes you dream of living the life of a château owner in the Loire Valley. This perfectly proportioned eighteenth-century home has beautiful formal rose and topiary gardens, complete with graceful statuary. It is surrounded by trees and vines that overlook the glistening Loire below. But the owner of Domaine aux Moines is not quite the chatelaine one might expect to be living in such a place. Cheerful and down-to-earth, Tessa Laroche is more hands-on winemaker than lady of the manor. In 2019 she changed her labels from 'Mme Laroche' to plain 'Tessa Laroche', at the same time removing the traditional royal crest of Anjou from her bottles, saying she was uncomfortable with the outdated image. During harvest she lives above the barrel cellar, 'so that I can smell if anything is wrong,' she says.

Tessa was born in Savennières, but did not grow up in these vineyards. It was not until 1981 that her parents acquired Domaine aux Moines and the covetable vineyard that comprises 12 of the 18 hectares of the Roche aux Moines appellation (the remainder is shared between 7 other producers). The earliest records of viticulture here date back to 1130 when the monks of the Abbey of St Nicolas of Angers planted vines on these rock-strewn hillsides. Tessa trained in Champagne, Aix-en-Provence and Bourgueil before coming back to work with her mother at the domaine in 2001. She now runs the place herself, with her small team.

Viticulture is organic and she practises biodynamics, although the domaine is not certified for the latter. The Roche aux Moines vineyard is exposed south and west, meaning that it is a warmer site compared to most of Savennières and ripens earlier. The soils are a mix of schists and granite, with slate, spilite and phtanite. Walking through the vineyard with Tessa is an education in the bewildering geology of the place, as she picks up a random handful of stones and identifies each one as a different type of rock.

Unlike many producers, who make multiple different cuvées, the Domaine aux Moines produces a single eponymous wine that reflects this unique site, although a smaller cuvée, Le Berceau des Fées, has been introduced recently, made from the fruit of vines under a decade old. This is a zesty, early-drinking style that has seen no oak and may be bottled without sulphur. As for the main wine, the grapes are pressed and the juice left to oxidize before fermentation ('to avoid oxidation in bottle,' says Tessa). Fermentation begins in stainless steel and continues in barrel, followed by a period of ageing. Malolactic fermentation is usual, and something she does not try to prevent. Commenting on the changes in style of Savennières wines, she says, 'It's a combination of factors: the climate, the people, and consumer tastes. Now the wines can be drunk earlier because the acidity is much lower.'

The Domaine aux Moines wine is full-bodied and ripe, but never overripe. These wines retain a vertical structure from their powerful mineral acidity, which ensures a long persistence in the mouth. Approachable earlier than some, these are generous and delicious wines.

After all the hard work of running the vineyard, when she does take time off Tessa says she likes to 'stay at home and do nothing'. She never feels the need to take a holiday. 'After all, I live in a paradise,' she says. It's hard to disagree.

Domaine du Closel

Savennières

www.savennieres-closel.com

Domaine du Closel is one of the old historic properties of Savennières with a beautiful château and park right in the middle of the village. This building, the Château des Vaults, has a history of wine growing that goes back as far as 1495. Evelyne de Pontbriand is currently in charge of the domaine, which has been in the hands of her family since the late nineteenth century. She continues a long line of female winemakers, taking over from her mother Michèle Bazin de Jessey in 2001. Evelyne and her mother have both been active in promoting and protecting their appellation, as presidents of the growers' association, a role Evelyne continues to hold today.

A former school teacher with passions for history, music and nature, Evelyne has not hesitated to take the domaine in a more natural

Evelyne de Pontbriand of Domaine du Closel

direction under her management, converting to organic and then biodynamic viticulture and adopting a more hands-off approach in the cellar. She has also been instrumental in developing tourism at the domaine, creating a vineyard walk for visitors that highlights the work of the domaine to increase biodiversity.

The estate produces three Savennières wines from its 15 hectares of vines. La Jalousie is from the domaine's younger vines. Here the grapes are picked early, when just starting to turn yellow, to produce an early-drinking wine that is fresh and fruity. This is the least expensive wine of the trio and is sound value. Les Caillardières comes from a parcel of older grapes that grow on a schist and sandstone plateau. Harvested later than La Jalousie grapes, this is a round wine with more complexity and a smooth palate. The top wine, Clos du Papillon, comes from one of the best sites in the appellation, named for its (passing) resemblance to the shape of a butterfly. It has excellent exposure and highly complex soils, with rhyolites and quartz as well as the ubiquitous schist that defines the appellation. This is always an interesting wine with layers of richly flavoured ripe fruit that sometimes verges on tropical, while retaining the appealing bitterness, restraint and fine acidity that makes these wines so enticing and ageworthy.

Evelyne de Pontbriand has done a great deal for the appellation and for her own domaine. She continues to be a driving force for change, motivating the other growers to join together to understand better their terroirs, how best to manage them, and how best to vinify their grapes. This is also a most welcoming domaine to visit. It is open year-round and visitors are able to taste at the château as well as enjoy the beautiful park and vineyards.

Domaine FL

Rochefort-sur-Loire

www.domainefl.com

Domaine FL is located outside Savennières and produces wines in both the Anjou and Savennières appellations. It is included in this section because it is one of the most significant producers in Savennières, with 24 hectares of land in the appellation. The modern winery-cum-visitor centre looks out across the domaine's vineyards from the other side of the Loire.

The company was founded in 2007 when Philippe Fournier, a local businessman with a background in telecommunications, took over the

Château de Chamboureau in Savennières and vineyards from Jo Pithon in Anjou to create today's domaine. The FL name stands for Fournier Longchamps, after Philippe's mother and father. Today the property is run by his son, Julien Fournier.

Five different Savennières wines are made, plus a wine from Roche aux Moines. Across the river, the range includes Anjous red and white, as well as sweet wines in Coteaux du Layon and Quarts de Chaume. Dry whites make up the great majority of production, and vinification methods are similar across the range. Fermentation begins in stainless steel before transferring the wines to oak barrels. The Savennières wines spend around 12 months in oak, and a small percentage of this (10–15 per cent) is new. One of the defining stylistic features of the FL white wines is that malolactic fermentation is systematically blocked. This makes for a style that is notably more acidic than most. The reason for this choice, says Julien, is to give his wines the ability to age. 'Malolactic removes the backbone of the wine,' he says.

The other important choice is to aim for homogeneous maturity, something that is not always easy with Chenin. The style of FL wines is certainly consistently good, however. When young, the wines can be quite closed and the oak more prominent. The basic Savennières comes from a blend of different sites. It is fragrant, ripe and polished, with a lick of oak. La Croix Picot, from the east of the appellation, is clean, crisp and fresh. La Frémine is more Burgundian in style, with more body and oak spice, and evidence of ageing on lees with regular *bâtonnage* to give added texture. The top cuvées are Chamboureau, from an old parcel surrounding the château of the same name in the heart of the appellation, and Le Parc, from an enclosed site where the vines grow on primary era volcanic rhyolite, phtanite and schist, topped with silty sand. Le Parc is a tremendous wine, deep, powerful and complex, with a strong mineral presence. The domaine's top dry wine is their Roche aux Moines, a riper style from this warmer site, balanced by FL's signature freshness. This and Le Parc are wines that can age for a decade.

The Coteaux du Layon and Quarts de Chaume Grand Cru wines from FL are both excellent, with richness and purity, but zero excess. The Anjou wines are unassuming and tasty – the red Anjou-Villages a supple infusion style and the white a good expression of the terroir, light on its feet with a bitter twist.

Open for visits and tastings (which have the added benefit of being bookable online) and with a summer programme of events, Domaine

FL offers a complete oenological experience in its tasteful modern facilities.

Domaine Laureau

Epiré, Savennières

www.domainelaureau.com

Damien Laureau has a background in agriculture rather than viticulture, although most would agree he has taken to the latter like a duck to water. From a family of cereal farmers outside Paris, he studied agriculture and spent time working with farmers in Romania before arriving in the Loire in 1999 at Le Clos Frémur, an organic farm close to Angers. Here he managed the estate with his wife Florence, also an agricultural engineer, cultivating fruit and vegetables alongside grapevines.

In 2006, they established their winery in Epiré, where they farm 6 hectares of Chenin Blanc, including a small parcel in the Roche aux Moines appellation. The estate is certified organic, but Damien is interested in pushing the boundaries further by adopting regenerative techniques for managing their vineyards. They make their own compost and minimize copper sulphate use (this being the main tool used to combat disease in organic viticulture), the aim being to develop the natural fertility of the soils. 'Even if we are organic, we are still killing life with sulphur and copper,' he says. He has been studying soil regeneration at Dr Elaine Ingham's SoilFoodWeb school since 2019, saying he is 'just at the beginning' of his research. The key, he says, is 'to observe what is happening' in the vineyard.

At harvest time, the grapes are sorted according to soil type to make five cuvées. Everything is harvested by hand, so this can be a very pressured time since Chenin can ripen very quickly. 'We need to do everything within four or five days,' explains Damien, 'It is becoming harder and harder to pick at the right level of maturity.' In the cellar, he takes his time with pressing the grapes and leaves the juice to settle before minimal-intervention fermentation in stainless steel tanks, barrels or amphorae, depending on the cuvée. The wines are bottled unfiltered.

L'Alliance is a wine from various sites, a blend of the fruit of young vines, fermented in stainless steel. For an entry-level cuvée, it is decidedly classy, with minerality and precision. Les Genêts comes from sandy soils that are usually slow to warm in spring. This spends a year in oak and 10 months in tank, during which time it is left to rest undisturbed on its lees. This is a richer mouthful with some creaminess, but still has fiery minerality and a well-defined structure that is highly appealing. Le Bel Ouvrage comes from a 70:30 blend of rhyolite and schist soils. The thin soils make for a very structured wine for the long term. It is long, powerful and complete – and unfailingly delicious. Damien's Roche-aux-Moines wine comes from a south-east facing half-hectare of land that gives a balanced and multidimensional wine. It ages in oak and sandstone jars. In 2019, the domaine introduced a new top cuvée, Champ Bourcier. Daniel recounts that this came about as a result of the hard frost of 2019, which reduced the crop significantly. From the same terroir as Le Bel Ouvrage, these grapes grow solely on rhyolite. The wine is aged for a year in oak and almost two years further in sandstone jars. 'I discovered the extra élevage gives more interest,' says Damien. This wine has lovely purity, depth, power and minerality. The extended ageing lends a textured mouthfeel, with gentle tannins.

Damien and Florence have created an impeccable range of wines that combine precision and mineral intensity with enormous drinkability; they never fail to please.

Eric Morgat

Savennières

www.ericmorgat.com

'Ten years ago, I knew exactly what to do; now I only have questions,' says Eric Morgat. He has been making wine in Savennières for nearly 30 years but says, 'Every year I say, I've never seen grapes like this before.' However, this is a man who makes some of the most in-demand wines in the area. He must be doing something right.

Eric's first vintage was in 1995, when he was fresh out of college, aged 25. Although from a local wine-growing family, he chose to go it alone, acquiring a small parcel of land in Savennières, and building his business from scratch – doing pretty much everything himself, as he continues to do today. This probably goes some way to explain his individualistic approach (and also why he's so hard to track down; he's a busy man).

He farms 6 hectares across several plots in Savennières and one plot across the Loire in the Layon Valley, rescuing land from abandonment, creating everything himself, growing and then reducing the size of his domaine to enable him to concentrate his energies fully on the best sites for the three wines he makes today.

Eric Morgat's Clos Serteaux vineyard, overlooking the Loire

Fidès and Clos Serteaux are the names of his Savennières wines. The largest cuvée is Fidès, which comes from a selection of plots that include one in the Roche aux Moines vineyard. Clos Serteaux is a 1-hectare single vineyard in a very beautiful setting surrounded by trees, directly overlooking the Loire. Left to abandon, it was patiently restored and replanted by Eric over the course of a decade, the first vintage of this cuvée being in 2015. It overlooks the striking Pierre Becherelle rock, a 450 million-year-old monolith painted by J. M. W. Turner in 1826. The third wine, Litus, is an Anjou blanc. These vines grow on the exceptional terroir of the Pierre-Bise vineyard in the Layon Valley. Here, there is a narrow vein of volcanic spilite that gives wines with a firm and mineral acid core.

Eric Morgat's wines are compelling, powerful expressions of Chenin grown on the hard and unforgiving schistous and volcanic soils of Anjou. 'The terroir imposes itself here,' he says, 'and everything happens very quickly.' What he is aiming for, he says, is transparency in his wines. Clearly a skilled winemaker, he claims to reinvent everything, every year to produce his wines. 'I have too much experience to say I understand,' he says, 'but I've abandoned the need to understand.' It's a confidence that shows in his wine, the quality of which speaks for itself.

Thibaud Boudignon

La Possonnière, Savennières

'The goal is not to make something good, but to make something great.' These are the words of Thibaud Boudignon, a man whose focus and commitment to his wines are as clear as his earlier dedication to becoming a judo champion, something he pursued full-time until the age of 23. The latter he relinquished when he realized this was an ambition he was not going to achieve. 'It taught me that a career goes quickly,' he says.

Thibaud wasted no time in applying himself to his new *métier*. Studies in his native Bordeaux were followed by *stages* at Châteaux Lafite, Olivier and Canon-la-Gaffelière in Bordeaux, then Yarra Valley and Burgundy before taking a job as winemaker at Château Soucherie in Anjou. In 2009 he started making wine for himself in his spare time, buying grapes and renting winery space. Wondering why so much of

Thibaud Boudignon with his prized Clos de la Hutte vines

the Savennières appellation was unplanted, he managed to acquire an abandoned plot in La Possonnière, in the west of the appellation, where he is now based. This purchase was the Clos de la Hutte, which he refers to as his *grand cru* site. After clearing it, he planted clones from ten different massal selections of Chenin, from Domaine Huet, Clos Rougeard, François Chidaine and Antoine Foucault, as well as a local specialist nursery. After leaving Château Soucherie in 2015, Thibaud's first vintage here was 2016.

The estate is managed biodynamically (although it is not certified) and there is not a leaf out of place. 'It's all in the detail,' says Thibaud, a comment one might apply to everything that is done here. The vines receive every care, and in Clos de la Hutte they are protected from spring frosts by the latest heated wire technology that ensures the temperature in the fruit area never falls below zero (see p. 15). A small area (20 ares) in the best part of this vineyard is planted to ungrafted vines whose fruit goes into the rare cuvée, Franc de Pied.

He has managed to acquire surrounding plots to make a total of 7.5 hectares of land here and across the river in the Layon Valley. In La Possonnière he has constructed a new winery close to the vineyards. Everything feels peaceful, organized, and at one with nature. Only a third of the Clos de La Hutte site is planted with vines, so biodiversity is high. There are sheep, beehives, and singing birds aplenty. Even at harvest, there is seemingly no drama here. After hand-picking and crushing the grapes – free from botrytis, which he describes as 'the enemy' – the juice is then stored at a low temperature until all the fruit is in. Only then are decisions made. 'We take our time,' says Thibaud, 'We are never in a rush.' In the winery, he has computer-controlled regulation of temperature and humidity. Most of the wines see some oak, but there is no recipe here, with various types and ages of wood and even different thicknesses of stave, all with the aim of achieving the precision he is looking for in his wines.

The domaine's Anjou blanc comes from vines in Saint-Lambert-du-Lattay. Thibaud describes it as an 'exotic wine with a saline palate, immediately enjoyable.' He's right. It has mouthwatering fruit and a crystalline purity that is found in all of these wines due, in part, to his avoidance of malolactic fermentation. The other wines are all in the Savennières appellation. Clos de Frémine, with a silky, floral mouthfeel, is all aromatics and lightness, finishing on a note of salty finesse. La Vigne Cendrée comes from a tiny, south-facing 0.5-hectare plot of purple schist, granite and clay. This is a hugely concentrated and complex wine, the initial silky texture hiding a latent power beneath. There is fiery acidity but beautiful ripeness, and savoury power from the higher proportion of clay in the soils. This is a wine for the long term. The Clos de la Hutte is a beautifully balanced wine from the still-young vines, with an endless finish in the mouth. It will only improve over the years as the vines age. With his Franc de Pied Clos de la Hutte cuvée, Thibaud says he is aiming to show 'the absolute transparency of minerality' of the site. It is matured in a glass 'globe' and indeed the minerality comes rushing out of the glass. This is the purest expression of Chenin imaginable, a positively spine-tingling experience. These are great wines from a thoughtful, committed and, yes, competitive winemaker, who is aiming for nothing less than the best.

Other producers to try

Château d'Epiré, Château de Plaisance (see Anjou), Domaine Belargus (see Anjou), Domaine des Baumard (see Anjou), Domaine Ogereau (see Anjou), Château Pierre-Bise (see Anjou), Loïc Mahé.

When in Savennières

The cobbled church square of Savennières is an ideal place to sit and have lunch in the sunshine and there is a charming local restaurant here that allows one to do just this. Frequented by winemakers and tourists alike, the aptly named Le Chenin (lechenin-savennieres.fr) offers a short, reliably delicious menu that includes fresh, original dishes to match the wines of the area. There is a good selection of Savennières wines, several by the glass. Another dining option, half-hidden next to the church in Epiré, is A Table Chez Mili (www.atablechezmili.fr), a homely place full of rustic appeal. Here, a single *prix fixe* menu is served at lunchtimes only, usually in the flowery garden if the weather permits. Wines are local, of course. In summer, the island of Béhuard is a lovely place for a walk along the banks of the Loire, a visit to the remarkable church built into the rock, or to visit one of the seasonal restaurants that look out over the river.

Coulée de Serrant, an icon of the Loire Valley

Vignobles de La Coulée de Serrant

Savennières

www.coulee-de-serrant.com

This is one of the defining domaines of Savennières, and the Coulée de Serrant its most famous single vineyard. Coulée de Serrant was awarded its own appellation in 2015, but it has been highly regarded long before this. In continuous production since the twelfth century this is, in the eyes of many people, the finest vineyard in Savennières. Historical writers and gastronomes have lauded its wine as among the best in France.

Almost as renowned as the vineyard is its owner, Nicolas Joly, who has become France's (and probably, the world's) most vocal advocate for biodynamics. He speaks with passion and energy about the importance of the vineyard above all else and was the founder member of the biodynamic producer association known as La Renaissance des Appellations (The Return to Terroir), which now numbers 230 members across the globe. He has also written several books on the subject and travels the world spreading his message with undimmed enthusiasm. Rather more low profile is his daughter, Virginie, who has joined her father in the day-to-day running of the estate.

The domaine produces three wines, one in each of the Savennières' appellations. As well as the monopoly flagship Le Clos de La Coulée de Serrant, they make Le Clos de la Bergerie in Savennières Roche Aux Moines, and Les Vieux Clos in Savennières. Nicolas remains the main spokesman and, one feels, the driving force behind the style of the wines today.

When one speaks with Nicolas, he is always keen to emphasize the importance of expressing the place – the terroir. These days few would dispute this, but when it comes to winemaking, there is less clarity. Questions about winemaking are inevitably answered by coming back to the source: the vineyard. In fact, Nicolas claims, 'A true wine should not have any oenology. If nature has given you good grapes, be sure to do nothing'. While it's not clear where he draws the line at defining 'oenology' – he does, after all, press the grapes and put them in a tank – the message is clearly that he favours the bare minimum of intervention in the cellar, while not adhering to any natural wine charter. For Nicolas, it seems, the focus is squarely on the vineyard, almost to the exclusion of all else.

The Coulée de Serrant vineyard is a steeply sloping, south-facing 7-hectare plot facing directly onto the Loire, which moderates the temperature and airflow in the vineyard. Autumn humidity rising from the river favours the development of botrytis in the grapes. The soils are very stony, of schist and sandstone, with areas of phtanite, the mother rock often very close to the surface of the vineyard. The soils drain easily, making this a very warm, dry environment for grape growing.The Jolys take the view that the grapes must be picked at high levels of ripeness in order to express the terroir, and this includes waiting for botrytis wherever possible. Their view is that waiting for a portion (though not all) of the harvest to become botrytized concentrates the berries and enhances the mineral expression of the terroir. This means, however, that the style of the wines is in marked contrast to the modern trend in Savennières today, where growers pick earlier to avoid botrytis at all costs. Tessa Laroche, owner of neighbouring Domaine aux Moines, tells me that she has a running gag with Nicolas, who picks his grapes in Roche-aux-Moines vineyard 15 days after her. When he sees her harvesting, he asks: 'The crémant is good?'; in reply, she asks after the progress of his sweet wines. The decision to wait for high ripeness in the berries gives wines with higher alcohol levels (15.5% abv is not uncommon here), usually with a few grams of residual sugar as well. And sometimes, the estate does produce wine in the *moelleux* style.

While there is no doubting the superlative nature of the vineyard sites owned by the Joly family (especially La Coulée de Serrant), the wines have been criticized in recent years for failing to live up to the pedigree of their origins. Specifically, the wines can show oxidation and other faults associated with a more natural approach to winemaking, although the Jolys point out that the deep yellow colour is from the ripeness of the Chenin grapes rather than from oxidation. That said, Nicolas does recommend opening the wines up to five days ahead of drinking them, which does suggest a certain tolerance for a more oxidative style. The Joly wines are to be celebrated for their individuality and adherence to a style that still has its committed followers around the globe. The Clos de la Coulée de Serrant is certainly a wine for the long term; a recent tasting of a 20-year-old vintage showed extraordinary concentration of honeyed, quince, floral, nutty and mineral fruit. Served as a *vin de contemplation* at the end of a meal, this felt very much an expression of a place and a person. Nicolas Joly sums up his philosophy thus: 'There are two types of beauty: natural beauty, which is never perfect but which has charm, and plastic surgery, which is perfect but has no charm. If a wine is impeccable, it has no charm, but if you let the place express itself, you end up with something that has the charm of the place.'

SAUMUR

Any visitor to Saumur will be struck by its pale, smooth-walled château overlooking the Loire. It advertises the beauty of the local tuffeau rock, a key contributing factor to the uniqueness of the wines from the region. There are countless cellars in the Saumurois. Many were dug as quarries to extract this malleable building material, which was in high demand for the construction of the Loire's castles, grand houses and even the humbler village dwellings close by. The caves themselves were also used as habitations, the remains of which can still be seen. Later, they were found to provide a suitable home for mushroom culture and, of course, for storing and ageing wine.

Wine production in Saumur and the surrounding villages is essentially divided between a small number of large sparkling wine companies and a large number of small vignerons who make still wine. Saumur fizz is big business: more than ten million bottles of bubbles are sold every year, accounting for more than 40 per cent of all the wine produced in the area. The abundance of cool, deep cellars provides not only a popular summer tourist attraction (in some places it's even possible to cycle through the caves), but also perfect conditions for the ageing of these traditional method sparkling wines. Much of the Loire's crémant is also aged in these cellars, which were first put into use for such purposes by the founder of Loire sparkling wine, Jean Baptiste Ackerman. This young Belgian spent a year in Champagne before arriving in Saumur for an apprenticeship at a wine company. He noticed that young Chenin Blanc wines, like the wines of Champagne, also had a tendency to 'foam', or re-ferment, giving them a natural sparkle. Armed with the knowledge he had acquired in Champagne, he applied the same techniques to the wines of Saumur to create wines in the same style, with a second fermentation in bottle.

Saumur Mousseux and Crémant de Loire wines can be made from a wide range of varieties, although the presence of the ubiquitous white grape, Chenin, is a must for white sparkling Saumur, where it accounts for at least 60 per cent of any blend. For sparkling Saumur rosés Cabernet Franc is the major variety, again at least 60 per cent of the *assemblage*. Once bottled after the primary fermentation, the wines undergo a second fermentation in the bottle and then age for a minimum of nine months. In practice, many crémants and Saumur sparkling wines will be aged for longer than this. The specialist sparkling producers make and sell Crémant de Loire alongside their Saumur wines, often bringing in grapes from across the whole Loire region. Unlike Crémant de Loire, the grapes for Saumur Mousseux wines can be harvested by machine, which lowers their cost of production but also potentially their quality. This, or perhaps the growing consumer appreciation for, and familiarity with, crémants, may account for the fact that Saumur wines are often less expensive than crémants.

The non-sparkling wines of greatest significance around Saumur come from Saumur-Champigny, an appellation created in the 1950s exclusively for reds that now accounts for sales of more than ten million bottles annually. These bright and juicy wines have long been fashionable in the bistros of Paris, and continue to be the red Loire wines most commonly poured in restaurants in France. The appellation is a roughly triangular area that covers nine communes immediately south of the Loire, close to Saumur. The vineyards are bordered to the north by the Loire, to the west by the Thouet (a Loire tributary) and to the south by the forest of Fontevraud. The key component of the soils here is the local tuffeau limestone, which is of unusual purity. The off-white colour of this noble stone takes on a stunning luminosity in the evenings here.

Saumur-Champigny growers are immensely proud of their tuffeau (it features heavily in all of their marketing materials), but with good reason. This magical stone can both absorb and drain water, as well as retain heat as needed. Wines of any hue grown on tuffeau tend to have marked acidity and freshness regardless of the vintage conditions. Rainfall is relatively low here and the Saumur microclimate one of the hottest in the Loire Valley, hence the name Champigny, which comes from the Latin '*campus ignis*', meaning 'field of fire'.

The Saumur-Champigny producers often make a distinction between what they call 'fruity' wines and 'complex' wines, the difference between the two typically being that the 'fruity' wines have less in the way of winemaking intervention and are intended for earlier consumption, whereas the so-called 'complex' wines have higher levels of tannin extraction, often with obvious use of oak and a higher price tag. Many of the vignerons produce wines in both styles. Of course, the ideal is for wines to be both fruity and complex, and the best producers manage to combine these attributes across their range.

Outside of the top echelon of producers, though, the early-drinking styles are often preferable because they display the primary, youthful expression of Cabernet Franc that promoted the rise of the appellation in the first place, and less of the harsh tannins that can come from over-enthusiastic extraction. Saumur-Champigny wines exude finesse and delicacy when handled well. The best producers are very, very careful with their winemaking, taking particular care not to over-squeeze their tannins, which can be fierce and rustic, and sometimes green. Younger vignerons tend to be more circumspect than their parents in this regard, recognizing that wine does not need to be impressively tannic in order to be worthy. Serving her 2009 Les Loups Noirs, the owner of Domaine de Nerleux, Amélie Néau, apologized for the tannin structure, still firm well over a decade later. The explanation? 'This was my first vintage; I did it with my father.' Other positive moves in Saumur-Champigny include a biodiversity programme in the vineyards. Since 2008 herbicides have been banned across the appellation and growers are required to grow grass in alternate rows among the vines.

Towards a new classification of the wines of Saumur

In his influential two-part work, *L'Anjou, ses vignes et ses vins* (1925), Paul Maisonneuve notes that writings from the mid-eighteenth century suggest there was no red wine at all in Anjou in the 1750s. For his part, he pays scant attention to the few red wines that existed at the start of the twentieth century, but dedicates many pages to the quality of the white wines of Saumur coming from named sites he refers to as '*crus*'. Today all of these '*crus*' are planted either exclusively or with a significant majority of black grapes to make red wine (or sometimes sparkling wine). Ironically, many of the top white wine growing locations identified by Maisonneuve are situated within the red-only Saumur-Champigny appellation. These include: Saint-Cyr-En-Bourg, Chacé, Varrains, Montsoreau, Turquant, Parnay, Souzay and Dampierre. Studying a map of Saumur-Champigny, one can see that many of these communes are right on the river, where the grapes grow on the top of tuffeau cliffs overlooking the Loire. In Maisonneuve's day, this area was known as La Côte des Blancs, and the road that ran through today's Saumur-Champigny the Route des Blancs. The local winemakers still refer to this road as such.

Tasting the few white wines made from these sites today, it is clear that Maisonneuve was not wrong: there are some stunning whites produced here. He gives a special mention to the commune of Brézé as producing extremely classy wines and goes on to praise the unique qualities of Saumur whites compared to the other wines of Anjou which are due, in his view, to the tuffeau subsoil. This confers upon them a terroir taste with, he says, 'remarkable qualities' and a lightness found nowhere else in the area. One might say just the same today, and the Saumur winemakers are currently putting together their case to create six *crus* that recognize once more these forgotten white wine terroirs. The official name for these selected sites is Dénominations Géographiques Complémentaires (see glossary), a first step in having terroir-specific sites recognized. For now, everyone is referring to them as '*crus*', a more readily understood terminology that is also less of a mouthful.

The most revered of these *crus* today, as in Maisonneuve's time, is Brézé, which gives wines of particular tension and verticality. Centred around the hill and castle of Brézé, the proposed *cru* extends (some say controversially) to parcels north and south where a similar terroir profile exists. The area formerly known as La Côte des Blancs is relabelled for the purposes of the *crus* as La Côte. This area covers most of the existing Saumur-Champigny appellation, so it is probably not considered politic to emphasize the fact that it might, after all, be a better terroir for whites than for reds. Either way, several growers are increasing their Chenin plantings in this area, recognizing the potential for high-quality whites. There are a further four *crus* to the south and west of the town of Saumur. The names are Brossay, Berrie, Courchamps and Puy-Notre-Dame. These, and the names of the other *crus*, are already appearing as adjuncts to the name of Saumur, in anticipation of future success with the INAO.

The reds of Saumur-Champigny are the best known of Anjou, and it is here that the greatest concentration of quality growers can be found, but good wines are being made all over the region. One just needs to know where to look to find talented winemakers who understand what to do with their Cabernet Franc grapes.

Cabernet Franc is also the main attraction for the two other red wine appellations in Saumur, Saumur Rouge and Saumur Puy-Notre-Dame. The latter is a small appellation of under 80 hectares, spanning a handful of communes south of Saumur, centred on the hilltop town of Le Puy Notre Dame. The basic Saumur appellation covers the entire Saumur area. Although a small quantity of Cabernet Sauvignon is permitted in each of these appellations, the relatively late-ripening grape remains of limited appeal this far north. Red Saumur may also contain a proportion of Pineau d'Aunis, a grape rarely seen outside the little-known Coteaux du Loir or Coteaux du Vendômois appellations further east. Compared to Saumur-Champigny and Saumur Puy-Notre-Dame, Saumur Rouge wines have higher permitted yields and a broader range of terroirs, with no requirement to plant grass or other green cover between the rows. The recently granted Saumur Puy-Notre-Dame appellation has the most stringent requirements of any of the red wines of Saumur, with significantly lower permitted yields (50 hectolitres per hectare compared with 57 for Saumur and Saumur-Champigny) and a higher density of planting. No chaptalization or use of oak chips is allowed and the wines must be aged until June of the year following the harvest before release. Only a few winemakers produce wines under the Puy-Notre-Dame label, and these wines tend to be more concentrated, structured and age-worthy than those of Saumur.

Although it was the whites for which this region was renowned in the past, today they account for only 10 per cent of the plantings of the Saumur area, compared to the comparable red wine figure of 55 per cent. 'Between 1970 and 1990, all the best white wine terroirs were pulled up to plant red wine grapes,' says Thierry Germain, owner of Domaine des Roches Neuves.

Saumur – key facts

- **AOC granted**: 1936
- **Hectares in production**: 852
- **Production volume**: 42,700 hectolitres/5.7 million bottles
- **Styles**: 43 per cent red, 47 per cent white, 10 per cent rosé
- **Grapes**: red – Cabernet Franc, Cabernet Sauvignon, Pineau d'Aunis; white – Chenin Blanc; rosé – Cabernet Franc, Cabernet Sauvignon, Pineau d'Aunis
- **Soils**: tuffeau limestone

Saumur Mousseux/Fines Bulles – key facts

- **AOC granted**: 1936
- **Hectares in production**: 1,319
- **Production volume**: 79,500 hectolitres/10.6 million bottles
- **Styles**: 100 per cent sparkling – 90 per cent white, 10 per cent rosé
- **Grapes**: Chenin Blanc, Chardonnay, Sauvignon Blanc, Cabernet Franc, Cabernet Sauvignon, Pineau d'Aunis, Gamay, Grolleau Gris, Grolleau Noir, Pinot Noir
- **Soils**: tuffeau and schists

Saumur Puy-Notre-Dame – key facts

- **AOC granted**: 2009
- **Hectares in production**: 41
- **Production volume**: 1,800 hectolitres/240,000 bottles
- **Styles**: 100 per cent red
- **Grapes**: Cabernet Franc, Cabernet Sauvignon, Pineau d'Aunis
- **Soils**: tuffeau with pockets of flint

Saumur-Champigny – key facts

- **AOC granted**: 1957
- **Hectares in production**: 1,375 hectares
- **Production volume**: 70,000 hectolitres/9.3 million bottles
- **Styles**: 100 per cent red
- **Grapes**: Cabernet Franc
- **Soils**: tuffeau
- **Villages**: nine – Chacé, Montsoreau, Parnay, Saint-Cyr-en-Bourg, Saumur, Souzay-Champigny, Turquant, Varrains, Dampierre-sur-Loire

Notable Saumur producers

Antoine Sanzay

Varrains

Chef or winemaker? Antoine Sanzay opted for the latter. His family had been growing grapes for generations, but chose to send the grapes to the local co-op rather than make wine. It was Antoine who decided to take the plunge and vinify his grapes. 'Otherwise,' he says, 'I'd have been a sommelier or a chef. I have a passion for flavours and cooking.' Like his ancestors, though, Sanzay says, 'I am a farmer first. I focus on the well-being of the plant in order to have the best grapes for the vintage.'

His first foray into winemaking was in 2002. Little by little, he reduced the amount of grapes sent elsewhere for vinification, until in 2013 he was entirely self-sufficient. The 12 hectares he owns are farmed organically and all are harvested by hand, with careful selection in the vineyard. In the cellar, he uses a range of materials – stainless steel and concrete tanks, sandstone eggs and wooden barrels – for fermentation and ageing. A self-confessed autodidact (he says he had 'a minimum' of education, so keen was he to get started), Antoine says he aims for 'minerality, density, depth – and, most of all, a lot of sincerity' in his wines.

It certainly feels as if a lot of love has gone into the young cuvées of La Paterne and Les Poyeux. These wines show flair and style, with silky structures and fine, ripe fruit. La Paterne is a blend of his different terroirs, free from any oak influence and macerated gently (with no interference) for 25 days. Les Poyeux is a concentrated wine with enveloping, savoury fruit and fine, persistent tannins. This is a wine that is hard to resist now, but should drink well for the next ten years. Sanzay also produces a small quantity of AOC Saumur Blanc, Les Essarts. Young Chenin Blanc vines grown on flinty soils give a wine of surprising intensity and complexity, with a chalky texture and lightly saline finish. This is a domaine whose wines are improving every year.

Arnaud Lambert

Saint-Just-sur-Dive

www.arnaud-lambert.com

The Lambert family domaine was founded by Arnaud's father, Yves, in 1996. Leaving behind a career in finance, he started his new life as a Saumur vigneron at 47 years of age. Although this was a change of direction for him, he was not unfamiliar with the area, as his father-in-law, Robert Néau, owned the significant Saumur-Champigny estate, Domaine de Nerleux, and Yves had managed to acquire some vineyard parcels in the years before he finally committed to wine. Arnaud joined his father in 2005 via a more conventional route, gaining experience in Bordeaux, Burgundy and Brittany. He pursued studies in winemaking and vineyard management, and even had a stint at cooperage.

Soon after Arnaud's arrival the pair began to convert the vineyards to organic viticulture. 'There was no focus on the soils before this,' says Arnaud. Today, under his sole ownership, everything is about the soils – and the other elements of terroir that make each vineyard unique. The red and white wines of the domaine are stylish, vineyard-specific cuvées that show precision and balance across the range. Reds are classified as Saumur or Saumur-Champigny. Arnaud comments, 'Here we are on limestone, which is white wine territory, so I make the reds more like white wines.' (This is a view apparently shared by several of the

Arnaud Lambert with a core of pure tuffeau rock

best growers of the region.) The grapes are picked relatively early ('If you wait for the seeds to be ripe you lose the aromas in the wine') and extractions are gentle, to preserve the fruit delicacy and avoid bitterness, to which he professes to be very sensitive.

The domaine's vineyards include multiple sites in the Brézé area, and Arnaud is one of the vignerons driving the proposal for recognition of six *crus* for the white wines of Saumur, including Brézé. White wines were historically recognized as being of superior quality to any others, and little else was grown here. The hill of Brézé was the most famous site in the area. The suggested *crus* would have the usual requirements for higher status, such as lower yields and longer ageing. 'We want to return to the eighteenth century,' says Arnaud. His whites are some of the best in the region. 'It's easier to make great white than red,' he says, 'because the reds really need to age.' He sees the future of the domaine as being more focused on whites, with the reds aged for longer before release. The whites have little to no malolactic fermentation, and are pure and vibrant. They typically age for a year in oak and are often not released for three to four years after the vintage, 'to show the minerality of the terroir'. But there is no recipe; Arnaud explains that the ageing will depend on the site exposure and soils. While all the wines of this domaine are of excellent quality, the Brézé whites La Rue and David are particular highlights. These show linear precision and minerality combined with detailed texture from lees ageing and complex, clear fruit, all put together with the lightest of touches. The reds also show restraint and elegance. Clos de L'Etoile, from vines planted in Brézé, has very light extraction ('this is white wine territory,' says Arnaud); it is bright, fresh, fruity and mineral-tinged. Moleton is a denser, ripe style, with a confit black cherry character and sooty tannins, all freshened by the beautiful acidity that comes with the terroir here.

Château de Targé

Parnay

www.chateaudetarge.fr

Château de Targé is a handsome turreted edifice that blends seamlessly into the tuffeau rock. This is the same rock that lies beneath the Saumur-Champigny vineyards above it – it's easy to see why the winery was built here. The caves for vinification and storage disappear into the rock face, and it is but a short walk up the (admittedly rather steep) slope to get to the vines of La Côte overlooking the Loire. Once in the vineyard, a 20-hectare section of the plateau, chimneys can be seen here and there. These openings, dug into the soil and carved from the rock beneath allowed harvesters in times past to drop their pickings through the hole and directly into the winery below.

Over the 350-plus years of ownership of the domaine by the Pisani-Ferry family the château has been used as a meeting place for politicians, as the family had many political connections. But today the focus is purely wine, under the management of Paul Pisani-Ferry, who has taken over the day-to-day management of the domaine from his father, although the latter remains involved as well. Paul has Loire and Saint-Emilion experience and also trained overseas at Raats in South Africa and Zuccardi in Argentina.

It is interesting to see the evolution of style between father and son. Paul jokes, 'Whenever my father receives visitors, they always end up buying different wines from when they see me.' Such is the power of suggestion and selling, of course – but also a sign of the change in preference between the older and younger generations. Paul has identified the suitability of his land for lighter styles that suit modern preferences, and the production of the domaine reflects this, alongside the more traditional offerings.

He has reined in extraction and oak in the domaine's red wines, moving towards more perfumed expressions of Cabernet Franc. A successful recent innovation was the Clos St Père, introduced for the 2020 vintage. This is a red wine with a very short maceration that is almost like a dark rosé in colour, and is designed to be served like one: 'I come home from work and have a glass straight from the fridge,' says Paul. It is an original and entirely gluggable reimagining of Saumur-Champigny wine. The Rosé de Targé is equally crisp and fresh, with a touch of oak. Another rosé, Méthode Ancestrale, is a dry, smartly packaged and fashionably pale best-seller. Paul is increasing the domaine's plantings of Chenin Blanc, recognizing (as are others) the suitability of this site above Parnay – formerly known as the Côte des Blancs – for white wine production.

As well as welcoming visitors throughout the summer, there is a beautiful bolthole for two for rent year-round. Reached by a tuffeau tunnel, it was formerly used for romantic trysts with the mistress of an amorous ancestor. It appears his wife drew the line at having his paramour actually in the château.

Clos de L'Ecotard

Courchamps

www.closdelecotard.com

The Clos de L'Ecotard domaine was created by Michel Chevré in 2010. No stranger to wine, he had worked as Thierry Germain's right-hand man at Domaine des Roches Neuves for 30 years before starting up on his own. His father had owned vines but, due to arguments among the six siblings, they ended up outside of the family. Michel had been buying land and planting vines from 1996 before making his own wine.

The domaine covers 15 hectares at the edge of the Saumur appellation, the last vestiges of Anjou blanc limestone before the start of the Anjou noir soils. There are few vines here; the rural landscape between the small towns of Doué-la-Fontaine and Montreuil-Bellay is one of rolling sunflower fields and cereals.

Since 2017, Michel has been working with his son Thibaud, who officially took over in 2024, although his father continues to work in a consultancy role.

The domaine makes three white wines that comprise the majority of production. La Haie Nardin is from old vines that grow on a sandy clay soil (the tuffeau is there, but a good three metres below ground), giving a creamy, tender Chenin expression with juicy freshness. The other two whites come from a 2-hectare south-facing slope close to the winery in Courchamps, planted by Michel at a density of 8,000 plants per hectare and ploughed by two horses, Fleur and Idée. These are the classic clay-limestone soils of Saumur. Thibaud explains: 'There is a lot of powdery limestone on top, then perhaps 40 centimetres of clay before the mother rock, close to the surface.' The high planting density gives smaller bunches, he tells me, which are all harvested by hand. The domaine's flagship wine, Clos de L'Ecotard, comes from the lower portion of this vineyard. It is a brilliant wine of great precision, with chalky texture and airy, mineral acidity – very much an expression of the place. The fermentation is unhurried, says Thibaud, and the wine spends a year in oak of which only a small proportion is new, so the wine feels very together – if contained – from its youth. Les Pentes comes from the upper part of the same vineyard, where the soil is a little thinner. This wine has more concentration and ages for two years in oak.

The other wines from the domaine are also worth exploring. The Clos de L'Ecotard Saumur Rouge is gorgeously open and enticingly floral. Fermented in fibreglass tanks, it ages partly in barrels and partly in sandstone jars. Thibaud describes the winemaking as 'infusion for 30 days with no extraction'. This is plump and gluggable, utterly fresh and lovely. The soft structure gives it immense drinkability. The domaine also produces two good traditional method sparkling wines, including Les Quarts Saint Vincent, which comes from 50-year-old vines. It ages on lees for 54 months and has zero dosage.

Clos Rougeard

Chacé

www.clos-rougeard.com

There have probably been more stories told and rumours spread about Clos Rougeard than any other domaine in the Loire. It's hardly surprising. There are many ingredients of a soap opera here – the death of a partner, the dramatic sale of the business, a family divided, the end of an era, and endless speculation about what the future might bring – but putting aside the family feuds and fights, claims and counter-claims for a moment, Clos Rougeard remains *the* reference for fine wine in the Loire Valley. At least at the time of writing – who knows what plot twists might follow? The status of Clos Rougeard has gone from an under-the-radar gem at the turn of the twenty-first century to a producer of cult wines that are all pre-sold before release. How did it happen?

Rewinding the story a few decades, brothers Charly and Nady were the eighth generation of Foucaults to manage the dozen hectares of family vines in and around the village of Chacé, in the Saumur and Saumur-Champigny appellations. The domaine goes back to the seventeenth century and these vineyards have been farmed by hand, without pesticides or herbicides, for as long as anyone can remember, and certainly before it was fashionable to do so. The brothers were unafraid to ignore the vagaries of the market and plough their own furrow (as it were), focusing their efforts on understanding the minutiae of their terroir and producing quality rather than quantity. As former technical director Jacques-Antoine Toublanc told me, 'People used to mock me in the 1990s for going to work for the people who let grass grow between their vines.' These days, of course, everybody is doing this – and the focus of the appellation is very much on quality. The Foucaults were simply ahead of the game with their old-fashioned ways. Winemaking and barrel ageing were as they had always been, with wines released at an unhurried pace a full four years after harvest.

After Charly's death in 2015, Nady took the controversial decision to sell the winery outside the family, to the Bouygues brothers (who own, among other things Château Montrose, a second classed growth in Bordeaux) in 2016.

The new owners have built a significant new winery and swish tasting room overlooking the Poyeux vineyard, in a style more suited to an icon wine (or perhaps a Bordeaux classed growth) than a humble Loire family domain. But the aim of all this is, according to new winemaker, Cyril Chirouze, to maintain the style and to respect the heritage of these wines. He says the new developments are intended simply to give flexibility in coping with climate change. No fewer than 17 new concrete fermentation eggs stand at the ready, each able to hold a hectare's worth of grapes, which will allow picking at optimal ripeness, whatever the conditions: 'If we want to keep the silky, elegant style of Clos Rougeard, we need to adapt,' says Cyril.

Other than the visibly impressive exterior and new tanks, little appears to have changed in the approach to winemaking thus far. Hand-picked grapes are hand-sorted – twice – and destemmed without

Clos Rougeard's Cyril Chirouze samples Le Bourg

crushing. After fermentation, the wines age in barrel in deep underground cellars for two years (now under artful lighting), and a further two years or more in bottle before release.

The three reds reflect the different soils of Saumur-Champigny. The domaine wine comes from 15 different plots across 5 hectares, some of which are flat, others steep and stony blends of clay and limestone. Les Poyeux is from 2 hectares in the *lieu-dit* of the same name, a gentle clay-limestone slope with vines of around 50 years old. Other wineries have land here too, and the name is not proprietorial to the domain, but the Rougeard plot is distinguished by a topsoil of wind-blown sand, which perhaps accounts for its diaphanous texture and streamlined elegance. Le Bourg comes from old vines centred on the village of Chacé. This soil is limestone, but with more clay (and none of the sand of Poyeux), giving denser, ripe, black-fruited wines that still retain the silkiness that characterizes all the reds of the domain. The single white wine, Brézé, is no less impressive, grown on clay-rich soils on the hill of the same name and aged for a year in oak. The 1.5-hectare plot faces north and west, giving a range of fruit profiles and complexity. The wine is rich, with flavours of *tarte tatin*, pastry, nuts and honey, held together by the chalky structure and verticality typical of Brézé, a site mentioned in historical texts as a *grand cru*-status site.

Hard to find but so easy to drink, Clos Rougeard wines are the epitome of Saumur-Champigny elegance and have long set the standard for the appellation. The Bouygues brothers seem committed to continuing the tradition of striving for the best.

Domaine de la Renière

Le Puy Notre Dame
www.domainedelareniere.com

Domaine de la Renière is an old farm that was established by the Gay family in the village of Le Puy Notre Dame in the 1600s. Several generations of the family have made wine here, and Thibault Masse is the first outsider to own the 16-hectare property. He took over in 2017 and his first vintage was in 2018. Thibault is from the area, and trained at the nearby viticultural lycée in Montreuil-Bellay. This was followed by further studies in Burgundy and Bordeaux and experience outside the Loire in Bordeaux and Australia, as well as learning the secrets of making great Chenin Blanc more locally with renowned producer, Romain Guiberteau.

Thibault takes a holistic approach. He is focused on building biodiversity in the vineyard and limiting inputs to the vines. The work is organic and harvest is done by hand. The focus of the estate is whites from Chenin Blanc. La Cerisaie comes from young Chenin vines. An early-drinking wine, it is clean, saline and crisp; it ages for 8 months in concrete tanks. The Signature (Thibault Masse) Cuvée is a beautiful wine from vines of more than 60 years of age. Honeyed and saline, with pear, quince and hints of exotic fruits, this wine ages for 18 months in amphorae and barriques.

Thibault also makes a bright, stainless steel-fermented Saumur red with short maceration and another red under the Saumur Puy-Notre-Dame appellation which comes from older vines. It has a short maceration but the required longer ageing period. As well as the excellent whites, other wines to discover here include a Cabernet Franc rosé, and a *pétillant naturel* from Chenin Blanc.

Domaine des Roches Neuves

Varrains

www.rochesneuves.com

'I'm a very anxious man,' says Thierry Germain, 'Every year it's as if I'm starting from scratch – I have no confidence in myself.' This must be quite a challenge, for he employs 14 people and owns 61 separate plots of vines over a 28-hectare area. Despite – or perhaps because of – this, he has made a considerable success of his wine business. And thankfully, now he is not alone. Since 2020 he has been working side-by-side with his son Louis, who has previously spent time in New Zealand working with biodynamic producer, James Millton. 'It's all the little details that make the quality of the wine,' remarks Thierry, happy to have his son to share the decision-making burden.

From a Bordeaux wine-growing family, Thierry moved here in 1991 with his wife Marie, taking over the domaine when it was less than half

Thierry Germain of Domaine des Roches Neuves

the size it is today and all planted to Cabernet Franc. It was chance that they ended up here, Thierry says – he had originally planned to go and make wine in Hungary instead. Throwing himself into this new venture, he expanded the vineyards over the years to include some prime sites of Chenin and further plantings of Cabernet Franc. He also converted the estate to organic and biodynamic viticulture from the late 1990s.

Today there are ten cuvées produced under the Saumur and Saumur-Champigny appellations, seven reds and three whites. The reds are characterized by their perfectly-ripe, crunchy fruit. Thierry is always one of the first to harvest. He was instrumental in getting the local authorities to ditch the requirement for growers to wait until they were told which date they could harvest (known as the *ban des vendanges*). 'The vision of Cabernet Franc has changed,' he says, 'people used to like sweet, ripe, high-alcohol styles like Bordeaux and wines from the south. It used to be difficult to get those things in the Loire, but now it's easy. The difficulty now is to search for the identity of the variety.' What he is searching for in his Cabernet Francs is what he refers to as '*le végétal*' – by which he does not mean vegetal, but more a notion of the vitality of the grape. 'People think of 'vegetal' as meaning pyrazines,' he says, 'but for me it means flowers and fruit – it's a positive thing. A living wine revolves around its acidity. The later you leave it to harvest, the greater the risk of losing that. At 14% Cabernet Franc loses its identity.'

This philosophy is very visible in the reds of Roches Neuves, which have appealing clarity and vibrant fruit even in the ripest of vintages, and enviable balance. 'The biggest problem is finding balance in the plant,' Thierry says, 'so we try to observe, and go in the direction of what the plant is trying to tell us.' Since 2006 there has been no green harvesting and no trimming the top of the vines. 'We let the plant go to the end of the vegetative cycle,' explains Thierry, 'and then the energy comes back down into the plant.' The fear is that by trimming the top of the vine during the growing season, the vine puts its energy into making more leaves instead of ripening its fruit. By rolling the top of the vine over the upper training wire, the vine finishes its cycle and ripens earlier and more evenly.

Harvest at Roches Neuves can take three weeks or more. Everything is picked by hand. The red wines spend a short time in oak (which is never new) but are bottled early. The whites are fermented in oak. The approach of the domaine inherently suits the terroirs of Saumur, where the chalky bedrock enhances the perception of acidity in the wines.

Terres Chaudes is a reliably delicious, scented red wine with floral notes from the vineyard of Les Poyeux. La Marginale is a wine of greater density, from deep clay soils over the limestone bedrock. It is concentrated and supple. Clos de L'Echelier is a 1.8-hectare site owned wholly by the domaine that produces mostly red wine, with a tiny portion of white made from an 80-year-old parcel of vines within it. 'It is the truest of my terroirs,' says Thierry. The red is a more linear style than Marginale, with an airy texture and elegant purity. The white is a remarkable wine, tightly wound with lemon-bright acidity and a distinct salinity. 'I'm looking for the crystalline side of Chenin,' says Thierry. Another red, Les Mémoires, comes from old vines that grow on clay, limestone and flint. Deceptively light in weight, it has a smoky intensity and fine-grained, powdery tannins that carry the chocolate and raspberry fruit to a persistent conclusion. Franc de Pied is a red from ungrafted vines that grow on a very sandy site (phylloxera, the predominant reason for grafting onto American rootstocks, does not thrive in sandy conditions). These grapes, from fairly young vines, nevertheless produce a wine of delicious intensity and purity. Another white, Clos Romans, comes from a vineyard within the Saumur-Champigny appellation (an appellation only for red wines) where the plants grow on a younger, harder limestone than most of the area. This gives compact wines of concentrated flavour, dense acidity and powerful minerality. It is an extraordinary wine, and unusual since almost all the land around it is planted with Cabernet Franc. 'This is the original Côte des Blancs,' says Thierry. 'People don't realize what they have under their feet – it's gold.'

Domaine des Sables Verts

Varrains

When I arrive at Domaine des Sables Verts, a smiling Caroline Malinge greets me holding the tiniest of babies. 'Let's go to the vineyard,' she says, passing the still-curled-up little human to her husband Hervé. There's no time for maternity leave here: Caroline and Hervé run their vineyards pretty much on their own and are in the vines every day, including the day before baby Jeanne was born. She was very obliging, says Caroline, arriving quickly and conveniently on time, and on a public holiday – a mere six days ago.

Domaine des Sables Verts is a bold undertaking for these first-time vignerons. It comprises 16 hectares in and around Saumur-Champigny, including two hectares of 70-year-old vines in the much sought-after

Caroline Malinge and her old vines in Les Poyeux

vineyard of Les Poyeux (known particularly for Clos Rougeard's cuvée of the same name). Visiting this vineyard, Caroline indicates the green sand that characterizes the topsoil of the upper part of this vineyard and from which the domaine takes its name.

The couple met while pursuing their wine studies together in Bordeaux. Caroline describes their acquisition of the domaine as a 'drunken dream' that by chance came true. She is from the north of France and Hervé from Vendée; after gaining some work experience in Bordeaux they decided to move closer to their roots. By chance, they heard of the rare availability of this domaine. The owner was retiring and happy to sell to a young couple just starting out. In France, vineyard transactions and financing often favour the young to encourage them to continue to farm the land. Hervé and Caroline continued a contract to send the grapes from eight of their hectares to the cooperative, and they work the best plots of the remainder for the domaine's wines. Their first vintage here was 2019.

They immediately began converting the vineyard to organic viticulture, listening to their neighbours to gain insights into how to manage Cabernet Franc, which makes up 15 of their 16 hectares. 'In Bordeaux they say it's a diva grape and difficult to manage in the vines and the cellar,' says Caroline, 'but Loire producers say it's easy here'. Their approach is to watch and learn and base their work 'on feelings rather than analysis,' and to leave some of their Bordeaux learning behind. 'We learned what not to do,' says Caroline, 'the first year we waited for the harvest, like they do in Bordeaux. Now we don't do that.'

At harvest time, all the work is done by hand. The careful approach in the cellar emphasizes fruit and avoids excessive extraction of tannins. The resulting wines feel fresh, smooth and modern. The domaine red cuvée is an early-drinking blend from different plots, fermented in stainless steel and bottled within a few months. It shows the matchless easy-drinking, juicy Cabernet Franc style at which this terroir excels in the right hands. Other red cuvées have some gentle infusion of the skins and then age in (non-new) barrels or, in the case of Les Poyeux, partly in amphorae. Les Poyeux is a fabulous wine, the old vines giving lovely depth of black fruit without any excessive concentration. It is pure elegance in a glass. Les Chaintres comes from grapes that grow on rockier, stickier soils that have more clay. This gives a deeper, more massive wine that sees more oak and feels a little more traditional in style. 'Older people love it,' Caroline confirms. There are also two barrel-fermented white wines from the domaine's Chenin Blanc. The domaine wine is bright, crisp and approachable, from early-ripening sandy soils, while Les Rouères is from a cooler site with a high proportion of clay. This is unusual terroir for Chenin, which can be difficult to ripen, but that produces a powerful, firm and saline wine that spends longer maturing in oak before release.

From a standing start, this fearless young couple is already doing great things. Their future looks very bright.

Domaine du Collier

Saumoussay

If there's a single lesson to be learned from visiting Domaine du Collier, it's the importance of taking one's time. Caroline and Antoine Foucault are not in a hurry to do anything, and it's a philosophy that seems to rub off when in their company. At their tranquil tuffeau home in the heart of the Saumur vineyards, it's easy to lose all sense of time until you realize you've spent five hours there doing nothing but wandering the cellars, tasting and talking in the company of two garrulous people who live and breathe their *métier*. The Foucaults never go on holiday, they tell me. 'Why would we ever want to leave here?' asks Antoine. In the moment, it's hard to disagree.

Antoine and Caroline Foucault take working from home to another level

The Foucault name carries special reverence in this part of the world. Antoine is the son of the late Charly Foucault, mentor to many a young Loire grower and one-half of the two Foucault brothers who owned the Loire Valley's most renowned family winery, Clos Rougeard. From a young age, Antoine worked alongside his father, tending the vines and learning in the winery. Antoine would have been the ninth generation of Foucaults to make wine from the Clos Rougeard vineyards, but this was not to be. Following the unexpected death of his father in 2015, his uncle sold the winery and vineyards lock, stock and barrel, to the Bouygues brothers in 2016.

The decision to dispose of the family land will always sting for Antoine, but the couple's own story promises a happier ending. They met at the Rougeard sorting table during the 1998 harvest. Caroline recalls the challenge of dating a Foucault: 'I realized that if I wanted to continue to see him, I had to enter his world. And that meant going to the vines to see him, because he was there every single day, just like his father.' Clos Rougeard was very much a family-run affair, but with Caroline on hand to help out, she was gradually accepted into the fold. 'She replaced my grandmother on the bottling line,' jokes Antoine (Caroline nods, 'It's true').

In 1999 the couple took the decision to branch out and create a winery of their own. The name is taken from the square, Place du Collier, in the tiny village of Chacé, which is the historic home of the Foucault family and Clos Rougeard. In 2015 they bought their current property, including a house directly above the cellars, just a few kilometres down the road from Chacé. 'Some people like to get away from their work at night,' says Antoine, 'but not us. We wanted to be close.' The cellars, 14 metres below ground, are vast, dwarfing the tiny production from the 8 hectares of vines they farm. They clearly love it. Antoine reiterates that, with a constant temperature of 12°C, 'everything is very slow here.' The wines spend two to three years in barriques, followed by a year in bottle.

Domaine du Collier's vast tuffeau cellar

The whites, grown on the nearby hill of Brézé, are of crystalline purity and freshness. Two Saumur appellation cuvées are made (the cuvée Domaine and La Charpentrie), both unerringly delicious, with the backbone and delicacy that characterize this special site. Winemaking is the same for both, but the La Charpentrie vines are over 100 years old. These are world-class wines of extraordinary depth that need time (that word, again) to reveal themselves. The young wines are tight and firm, with salty and mineral intensity and a fine, chalky texture. Glimpses of citrus fruits, fennel, *garrigue* herbs, hazelnuts and acacia honey promise a long and fascinating evolution.

The domaine produces a single red in the Saumur-Champigny appellation, from a hectare of 40-year-old vines in Poyeux. A relatively recent acquisition (from the 2018 vintage), it is intensely violet-scented, concentrated, long, and exceptionally easy to drink. The long-standing reputation of the domaine's reds comes from two cuvées grown on the same slopes of Brézé as the whites (Saumur appellation). Antoine points out, 'It's a white wine terroir, so you need to be very careful with reds to avoid astringent tannins.' Again, time seems to be the answer. La Ripaille shows a mass of powdery tannins, somewhat accentuated by the naturally high acidity that comes from this site. Combine this structure with all the perfumed, lifted elegance of fine Cabernet Franc in the shape of cloves, violets and cassis, and the result is a stunning wine. The red La Charpentrie shows extraordinary silk-and-velvet tannins cosseting fruit that, depending on the vintage, includes a magical and evolving blend of cherries, chocolate, tarragon and fennel seeds, crushed strawberries, truffles, cloves, dried herbs, cassis, and more. These are truly gorgeous wines, concentrated without a trace of heaviness, approachable yet deeply complex.

Domaine du Collier is artisanal wine production at its finest, a result of long hours of attentive work in the vineyard, and long years of patience in the cellar. At the risk of sounding sentimental, one can't help feeling there's a healthy dose of love in these wines, too. How else could they be so very good? Antoine smiles. 'As my father always said, '*Ce n'est pas "que" du vin*' – it's not 'just' wine.

Domaine Guiberteau

Saint-Just-sur-Dive

www.domaineguiberteau.fr

It's not a bad thing to reach the end of the Impasse du Cabernet when

Romain Guiberteau enjoys a drop of his Brézé wine over lunch

the road leads you into the cellars of Domaine Guiberteau, one of the most exciting domaines in Saumur today.

The Guiberteau family have been in the little village of Saint-Just-sur-Dive since 1934, when they set up the domaine. Romain Guiberteau has been running the 17-hectare estate since 1996. He makes red and white wines, in roughly equal proportions, from Cabernet Franc and Chenin Blanc. All of the wines are in the Saumur appellation.

Romain makes four reds and four whites. 'I aim never to release an average wine,' he says. For the reds, all the grapes are destemmed to avoid accentuating any of the green tannins to which Cabernet Franc can be prone. After fermentation at carefully controlled temperatures, the wines spend between four and eight days on skins. The aim here is 'to work towards finesse and elegance,' I am told. 'We like Pinot, Gamay, northern Rhône styles,' says Romain. Over the years, the maceration times have become shorter and the wines more delicate in style.

Les Moulins is an early-bottled cuvée that shows exactly this style. Light in colour, with a floral scent, it has delicate, gentle fruit and very juicy acidity. It's immediate and delicious. The estate red ages for a year in wood and a further year in bottle. It is from a blend of sites, mostly from vines between 50 and 70 years old. The fruit comes for the most part from Brézé, which gives the wine very fresh acidity. It has quite intense black fruit appeal and a silky texture. Les Arboises is also from Brézé fruit. The wine ages for two years in barrels, 50 per cent of which are new. Oaky when young, it has fruit sweetness, fine-grained tannins and balanced acidity. It needs time to evolve, as does Les Chapaudaises, which also ages in oak, giving it a tarry edge.

The reds are certainly good, but it is the pristine whites that are the *grands vins* here. The early-bottled Domaine cuvée shows light, pineapple fruit and scents of white flowers. It is easy, juicy, early-drinking elegance. The Clos de Guichaux white comes from the same tuffeau terroir as Les Chapaudaises. It ages for up to a year in old oak, giving the linear, chalky fruit a slight roundness that takes nothing away from the natural tension of the site. It's long, dry and delicious. Le Bourg Brézé is made from the fruit of 80-year-old vines and spends a year in oak. It has the rapier-like acidity characteristic of Brézé combined with a delicious ripeness to the fruit and a light veil of smoky oak. It is long, mineral, savoury and lightly saline, and feels almost weightless in the mouth. A further white, Clos des Carmes, also from Brézé, spends two years in oak, which imparts a layer of cinnamon and clove spice to the wine. Malolactic fermentation is blocked for the white wines, which emphasizes their natural verticality.

Domaine Théo Blet

Courchamps

Théo Blet's family is not new to grape growing but, until recently, the family's produce was sent to the local cooperative. 'My grandfather made wine,' Théo says, 'but when the co-op was set up in the 1970s, it made life easier to send the grapes there.' Grapes are just a part of the business here: the family has 15 hectares of vines and 120 hectares of cereals – wheat, corn, quinoa and sunflowers. The vines, though, take up 70 per cent of Théo's time, he says. Most of the grapes are still destined for the co-op, and the domaine is currently only sourcing grapes from around a hectare of the vines, Théo selecting what he thinks will be the best plots. So far, the domaine has one red and one white wine.

The first vintage was 2021. His white wine, Le Peyanne, is an easy, fresh and citrussy Chenin, bright as a button, with featherlight finesse. It has a lightly salty finish with grapefruit and clementine tang. The Cabernet Franc red, Le Corbin, has an extraordinarily scented nose of cassis, crushed strawberry and cherry fruit. The crunchy fruit bounds out of the glass. It's youthful and joyous. The Saumur tuffeau is still present in these clay and limestone soils to the south-west of Saumur, giving the wines acidity and mineral interest. Théo aims to grow the domaine to 5 hectares in the next few years, by continuing to identify the best sites for his wines.

Other Saumur producers to try

Brendan Stater-West, Bouvet Ladubay (sparkling), Château de Bellevue, Château de Chaintres, Château de Fosse Sèche, Château de Villeneuve, Château du Hureau, Château Yvonne, Domaine de la Renière, Domaine de L'Austral, Domaine de Nerleux, Domaine de Rocheville, Domaine Filliatreau, ExMuros, Fabien Duveau, Jonathan Maunoury, Le Pas Saint-Martin, Louis de Grenelle (sparkling).

When in Saumur and the surrounding area

Saumur is a popular tourist centre with plenty of options for wine tastings, visits, cycle hire and restaurants. During summer there are many wine-related activities: check out the tourist office (www.ot-saumur.fr), or the Maison des Vins d'Anjou et de Saumur (www.vins-de-saumur.com/visiter/la-maison-des-vins-anjou-saumur) for details. Both can be found on the quayside in Saumur.

In the centre of the old town, Bistrot de la Place (www.bistrotdelaplace-saumur.com) serves a wide range of traditional French fare and has an excellent, organic-focused wine list that features many hard-to-find gems. There are 40 by-the-glass options. It's popular among the locals and the wines are also available to take away at very reasonable prices. There are few nicer places to soak up summer in the Loire Valley than beneath the awning of the terrace on a Saturday lunchtime, surrounded by the sights, sounds and smells of a classic French market. Another charming option further along the quayside on a quiet street is Le Pot de Lapin (www.lepotdelapin-saumur.com). In the traditional beamed dining room, a giant blackboard displays a changing selection of home-cooked delights, served with a smile and an interesting selection of wines.

In Montsoreau, one of the Loire's prettiest riverside villages, Ververt is a great choice for elegant, fresh cuisine with a strong emphasis on vegetables and market produce (www.ververt.com). It also features an excellent wine list. A relatively recent opening, it is extremely popular, so needs to be booked ahead, especially on the second Sunday of the month, when they have a special five-course lunch menu to coincide with the local *brocante*. A little further along the river from Montsoreau is a cute wine bar and shop, Enfin du Vin (www.enfinduvin.net), offering wine tastings and sharing platters of cheese and charcuterie, along with music soirées in summer.

In the quiet little wine village of Le Puy Notre Dame is Le Puy à Vins (www.lepuyavins.fr), a relaxed restaurant and wine shop specializing in organic and natural wines. Situated opposite the magnificent church, diners can choose from a short menu of local, freshly prepared home cooking.

HAUT-POITOU

The Haut-Poitou appellation lies some way south in the Vienne region, on a limestone plateau to the north-west of Poitiers. It has been elevated to appellation contrôlée status relatively recently, in 2011. The wines are for the most part light and floral varietal white wines from Sauvignon Blanc. There are also palatable, light reds and rosés from Cabernet Franc-dominant blends. The climate is sunnier than most grape-growing spots in the Loire and the vines can produce reliable yields of up to 67 hectolitres per hectare for the whites and 60 for the reds and rosés. These are wines for early-drinking refreshment. The wines of La Tour Beaumont (www.domainelatourbeaumont.fr) are good examples of the modern, fresh styles that can be found here.

Haut-Poitou – key facts

- **AOC granted**: 2011
- **Hectares in production**: 117
- **Production volume**: 6,500 hectolitres/870,000 bottles
- **Styles**: 83% white, 12% red, 5% rosé
- **Grapes**: whites – Sauvignon Blanc, Sauvignon Gris, reds – Cabernet Franc, Gamay, Merlot, Pinot Noir, rosés – Cabernet Franc, Pinot Noir, Gamay
- **Soils**: clay-limestone

6

TOURAINE: THE HEART OF THE LOIRE

Touraine can probably boast more châteaux per square kilometre than anywhere on earth, but as well as being at the geographical centre of the UNESCO-classified Loire, it is also at its viticultural heart. Second only to neighbouring Anjou in size, it includes some of the best-known appellations of the Loire, as well as some of its most obscure.

Touraine begins immediately to the east of Saumur. It marks the crossroads between a climate that is more ocean-influenced to the west and one that is increasingly continental towards the east, where the temperature differences between day and night are more marked. Rainfall is also significantly higher in the east than the west (650 millimetres compared to 550). As is always the case in the Loire Valley, the growing environment is affected by the presence of rivers. Here, the Vienne, Indre, Cher and Loir are all important influences on the terroir of the vineyards planted in the appellations that surround them, as well as those of the Loire itself. The rivers account for differences in temperature, wind flow and soil composition and formation.

Touraine is the Loire's most important region for red wine, with a concentration of three appellations – Chinon, Bourgueil and Saint-Nicolas-de-Bourgueil – devoted almost exclusively to producing red wines from Cabernet Franc. For white wine grapes, the most significant production is Sauvignon Blanc, which covers a large area in the east of the region. Touraine is also important for Chenin Blanc, especially the dry, off-dry and sparkling styles for which the uneasy neighbours Vouvray and Montlouis are famous. Elsewhere in the region there are

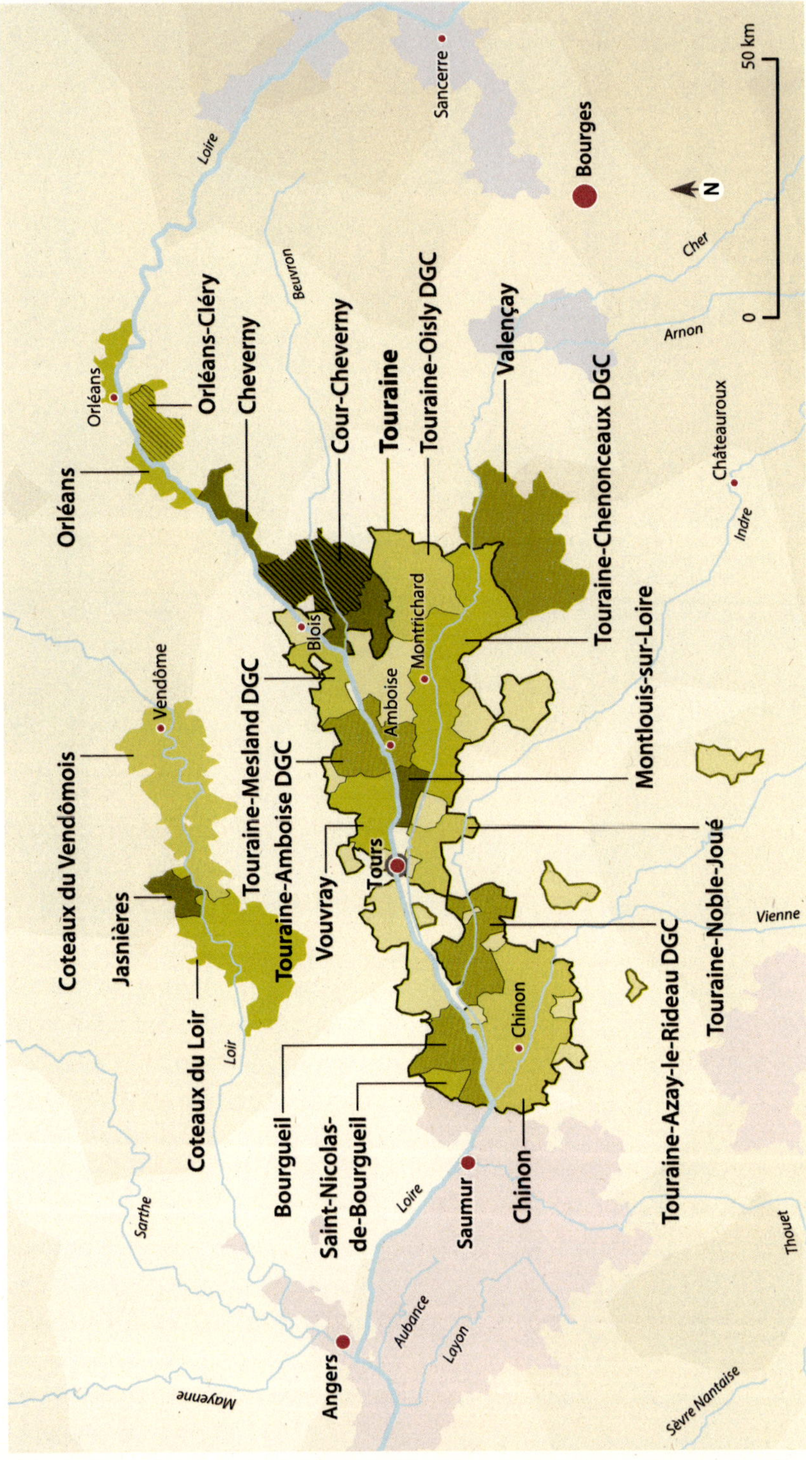

Map 8: Touraine

less well-known gems to discover, such as Pineau d'Aunis in the sleepy Coteaux du Loir region, fierce Romorantin wines in Cour-Cheverny, or the gentle Chenins of Jasnières.

Touraine – key facts

- **Hectares in production**: 12,900 (23 per cent of Loire total)
- **Annual production volume**: 574,000 hectolitres/77 million bottles (21 per cent of Loire output)
- **Styles**: 43 per cent red, 38 per cent white, 13 per cent sparkling, 6 per cent rosé
- **Most important grapes**: Sauvignon Blanc, Cabernet Franc, Chenin Blanc

Touraine appellations

- Bourgueil
- Cheverny
- Chinon
- Coteaux du Loir
- Coteaux du Vendômois
- Cour-Cheverny
- Jasnières
- Montlouis-sur-Loire
- Orléans
- Orléans-Cléry
- Saint-Nicolas-de-Bourgueil
- Touraine
- Touraine Noble-Joué
- Valençay
- Vouvray

VOUVRAY AND MONTLOUIS-SUR-LOIRE

Vouvray

One of the most famous appellations in the Loire Valley and – no coincidence, perhaps – one of the easiest for Anglophones to pronounce, Vouvray is the rounded, off-dry expression of Chenin in many people's minds when they think of Loire whites made from this variety. Although the appellation is best known for wines with at least some sweetness, today there is an increasing focus on dry wines. It may come as a surprise to learn that the majority of the production here is destined for sparkling wine, but this has long been the mainstay of Vouvray production, even if it is the still wines that offer the greatest interest for the wine lover.

In Vouvray, the influence of the ocean is less marked compared to vineyards further west, and the climate is semi-continental. Temperatures remain high late into the season, allowing grapes to become ripe enough to make the sweet wines for which Vouvray is historically renowned. Around three times a decade, botrytis develops to a level that allows sweet wines to be made from grapes affected by noble rot. In some vintages, by making multiple passes in the vineyard during harvest, it is perfectly possible for growers to make wines in all styles, from sparkling to dry to sweet.

Vouvray – key facts

- **AOC granted**: 1936
- **Hectares in production**: 2,212
- **Annual production volume**: 107,000 hectolitres/14.3 million bottles
- **Styles**: white – 52 per cent sparkling, 48 per cent still, all sweetness levels
- **Grapes**: Chenin Blanc, up to 5 per cent Orbois
- **Soils**: limestone base, with clay, sand and flint topsoils

Situated just to the east of the city of Tours, some of the vineyards of Vouvray actually touch the outer limits of the city. The rest lie eastwards across seven communes (Vouvray, Rochecorbon, Parçay-Meslay, Vernou-sur-Brenne, Noizay, Chançay and Reugny), directly above the Loire and extending back from it along a series of valleys. The finest sites are considered to be on the *première côte*, closest to the river. These vineyards face south and have thin topsoil over deep tuffeau limestone, the chalky white rock face easily visible from the road that runs alongside the Loire. This limestone bedrock is a common feature of all Vouvray vineyards, but further back from the river the overlying soils tend to be deeper, with more clay, along with some sand and flint. Further inland, there is also some silt on the plateaus. Slopes are gentle in Vouvray, and it can be difficult to distinguish where a slope ends and a plateau begins, but the change in soil type is key. In fact, soil is of such primordial importance to the vignerons here that they have collectively spent 200,000 euros in commissioning the French agricultural research institute, INRAE, to map the detail of the region's geology in an attempt to distinguish and market the distinctive qualities of their vineyards and *lieux-dits*.

Soil is not everything, however. Exposure is critical to the ripening of the crop. While a southern exposure was always thought in the past to be the most beneficial for ensuring optimum ripeness, the increase in warmer Loire vintages has meant that cooler inland sites have become sources of excellent, balanced dry wines that producers with more 'favourable' south-facing vineyards have struggled to produce.

Sparkling wines are made using the traditional (Champagne) method although, unlike Crémant de Loire or Champagne, the grapes may be harvested by machine, and indeed most of them are. In fact, most of the still wines of Vouvray are also harvested by machine, which some see as a quality problem for the appellation. According to one well-known grower, fewer than 30 of the 150 or so producers here harvest by hand. There are advantages and disadvantages to machine harvesting, of course. It is not possible to be selective when harvesting by machine, so the quality and uniformity of the crop may suffer, and by default there is no possibility of making a range of cuvées at different ripeness levels within a single parcel if a grower chooses the mechanical option. There is also the risk that the berries oxidize when they are shaken from the vine, ripping the skins. On the other hand, there are cost savings to consider. Labour is both expensive and increasingly hard to come by. Mechanization also equals speed, an advantage in vintages like 2020 when the grapes ripened so quickly that growers were left scrambling to harvest before the potential alcohol levels in the grapes soared.

One of the issues *du jour* in Vouvray is the common perception among growers that there is little appetite these days for wines made in the more traditional off-dry style. There is therefore an increasing trend to focus their production on wines that are completely dry. In some years, such as the lean, cool vintage of 2021, this approach was appropriate, especially in the cooler parts of the appellation which are naturally better suited to this style. But in Vouvray the vintage is, and always has been, key: in warmer vintages it can be a real challenge to manage the high levels of sugar that Chenin grapes naturally accumulate and to make a dry wine that is balanced. Often the result in these instances is that growers determined to stick to dry styles end up making wines with higher levels of alcohol than they would like and the wines can be out of balance. In the past, these producers would have routinely made *demi-sec* or sweeter styles with grapes like these.

One issue from a consumer perspective is that it is difficult to know how sweet the wine is going to be when buying a bottle of Vouvray simply by looking at the label. Although there are quite reasonable guidelines for growers regarding what constitutes *sec*, *demi-sec* or *moelleux*, the only indication that you are likely to see on a label is the obligatory 'Sec' if the wine is dry – except that, confusingly, sometimes this isn't there either. More often than not, there is no indication at all about the wine's level of sweetness. This is particularly troublesome in the *demi-sec* category where many of the wines sit. Growers would clearly prefer to be selling a dry wine because they believe this is what the consumer wants, but they don't have a dry wine to sell. Some have created special cuvées where there is no mention of what the sweetness level is, seemingly in the hope that their clients will become loyal to the brand and focus less on the level of sweetness. It is true that Chenin often benefits from some residual sugar to balance the high natural acidity in the grapes. But it is arguably always helpful to give your customers some idea of how sweet a wine is going to be before they open the bottle.

Notably, these are not problems you see among top growers, who are more inclined to let the conditions at vintage dictate which wines they make. Domaine Huet's Sarah Hwang is in no doubt about the value of sweeter Vouvray. '*Demi-sec* and *moelleux* are the heart and soul of what Vouvray represents,' she says. 'It's what you can't get elsewhere.' Selling the wines doesn't seem to be a problem if they are good enough to start with. It is perhaps a vicious circle: if the wines have sufficient renown, the prices are high enough for winegrowers to be able to afford the TLC in the vineyard that will enable such wines to be made. But it is also true that fashions change. Wine growers are at the mercy of the market and need to adapt to it. It is perhaps a leap of faith for growers to continue to do what, in the end, Vouvray has always done best: produce wines that reflect the vintage conditions of each year. That, in itself, should be something to celebrate.

Notable Vouvray producers

Clos de la Meslerie

Vernou-sur-Brenne

www.lameslerie.com

For New Yorker Peter Hahn, the decision to come to France and make wine was, he says, 'all about the lifestyle'. As a fellow city dweller, it's hard to argue that his blue-shuttered eighteenth-century countryside home, surrounded by 4 hectares of ancient vines, is anything but idyllic. But it's taken him a while to get to this point, having left his career as a financial consultant back home. He acquired the domaine in 2002 and his first vintage was not until 2008. At the time of purchase the land was rented out, being farmed conventionally, and for high-volume production by another grower.

Peter Hahn pruning at Clos de la Meslerie

Peter confesses that before he launched himself into this project he'd never knowingly even tasted Vouvray, but he was keen to learn. He went back to school to train himself up in viticulture and oenology at the Lycée Agricole in Amboise, and took guidance from old hand and mentor to many a young vigneron, Vincent Carême. Finally, he took back the land, aiming to breathe new, more natural, life into the old vines. He now works organically with a touch of biodynamic spraying and has a no-till approach to viticulture. Together with his partner Juliette, they do everything in the vineyard themselves. 'We wanted everything on site,' says Peter, and this is exactly what he has got. The vines grow on all four sides of the property. They are up to a century in age, and Peter only replaces dead plants using the old technique of layering, or '*marcottage*'. This entails burying a cane of the neighbouring plant in the ground so that a new plant grows in place of the dead one. It's not uncommon to see this in Vouvray. It has the advantage over grafted vines that there is a high success rate for the new plant, but as it is ungrafted it comes with the associated risk of phylloxera.

Peter's approach to winemaking is to make a single wine every year from all of his grapes, rather than the more common fashion for making multiple, site-specific cuvées. He uses a 100-year-old hydraulic basket press for his hand-picked grapes. After settling, the juice ferments in 225- and 228-litre barrels with indigenous yeasts, staying in the barrels for 12 months. Here, the wines rest on their lees and are infused with the strains of Verdi opera playing in the cellar – or whatever the couple decides is the music of the year this year.

Peter is not aiming for any specific style in his wines. He says, 'I like to see what the earth gives us. Every year it's different.' This means that the wine's sweetness level will vary depending on the conditions during the year and at vintage, and whether the alcoholic fermentation stops before the sugar has all been consumed by the yeasts. It is impossible to predict, therefore, what style of wine will emerge from the juice that goes into barrel, which is perhaps a true reflection of the historic production style of Vouvray. With a few exceptions, the wines have almost all had some residual sweetness since 2008 – again, perhaps a reflection of what Vouvray does best. But regardless of the style, they uniformly show lovely balance, with honeyed fruit and chalky texture. Peter also produces an excellent traditional method sparkling Brut Nature from wine that has already aged for 12 months in barrel and then undergoes a second fermentation in bottle with a further two years' ageing on lees. This is dry, full-bodied, yeasty and really good.

In a good year, says Peter, they will make 10,000 bottles. It doesn't seem a lot for the amount of hard work that goes into producing each one. Is he pleased to have made the switch from city ease to a life of country toil? 'I have no regrets,' he says, and gesturing to his land, adds, 'every morning I wake up to this.'

Domaine du Clos Naudin

Vouvray

The quality pyramid for Vouvray wine narrows sharply at the top, but Domaine du Clos Naudin is indisputably at the pinnacle, producing a matchless range of wines year after year. Founded in 1923 by Armand Foreau, the domaine consists of 12 hectares of the best of the Vouvray vineyard in the heart of the appellation. Since 1983 it has been run by Armand's grandson, Philippe Foreau, and Philippe's son Vincent is now making the wines with his father. Vincent exudes the confidence that comes from knowing you are top of the tree. 'It's not complicated,' he says, with a shrug.

I am not convinced. Surely there have to be some secrets to these wines, or everybody would be making them? Or maybe it's just the terroir.

The work in the vineyard is straightforward, and the fruit is harvested by hand. During the season the soil is worked from mid-March to the end of July, but otherwise remains untouched. In the vineyard, there are no pesticides or herbicides, but the domaine is not organic. 'We don't do logos,' says Vincent, 'and we don't believe in biodynamics. We don't feel the need to justify what we do; you'll find the truth in the glass.' Nor do they plant cover crops between the rows: 'It lowers acidity and raises the pH,' Vincent tells me. Of their 12 hectares, only around ten are in production. Once the vines in a plot reach 50 years old, those that die are not replaced; eventually the vines are pulled up en masse and the soil is left to rest for four or five years, during which time they sow cover crops to nourish the soil. Average yields of 38 hectolitres per hectare are well below the appellation's permitted level of 52.

Unlike many domaines, at Clos Naudin (which is, incidentally, not a *clos*, meaning walled vineyard) there is no interest in producing a range of cuvées from different plots. It's refreshingly simple: they make sparkling, *sec*, *demi-sec*, *moelleux* and *moelleux réserve* (and, very occasionally, a further sweet wine, Goutte d'Or). All the soils are basically the same,' says Vincent, simply, 'and the furthest is only a ten-minute walk from the cellar. We don't do analyses of the soil; we're anti-analysis. Instead, we look at the vines and taste the wines.' It's a defiantly old-fashioned approach, but it's maybe hard to argue with the 'if it ain't broke, don't fix it' mentality and, as Vincent points out, 'We don't work with numbers because we have something unique: our palates.' It is this savoir-faire and trust in working with nature that perhaps makes all the difference. It's a philosophy that means it is only at harvest time that the Foreaus decide which styles of wine they will make. In the warm 2018 vintage, for example, you can find many dry Vouvray wines from other domaines that taste unbalanced. Here, they made no dry wines in 2018. Vincent explains, 'In 2018 the grapes were not ripe until they'd reached 15% to 16% potential alcohol. We could discern this by eye and by tasting. So it was impossible to make sparkling wine or dry wine in 2018.' These are decisions that are taken in the vineyard, 'bunch by bunch, plant by plant'.

In the cellar, the vinification is – predictably – 'simple' and maturation, all in older wood, is short. The wines are bottled in the spring. 'Chenin is a sensitive variety,' Vincent says, 'we don't want to tire it.'

A tasting in the Foreau cellar is an enlightening and, at times, moving experience. These are wines of depth and character, elegance, freshness and refinement; some are simply breathtaking. All – no matter the level of sweetness – have a feeling of lightness on the finish that invites another sip … and another. Vincent moves deftly from one vintage to another, backwards and forwards to compare and contrast the wines. The 2021 Sec he declares 'strict – we need to wait ten years,' following it with a 2010 to compare 'another cold vintage'. After more than a decade in bottle this was singing, with flinty tangerine fruit and fabulous balance. 'It needs a carpaccio of St Jacques,' said Vincent, and thus began a series of passionate descriptions of all the dishes and flavours to match each of the many wines he poured: 'langoustines, gambas … sweetbreads, morels … here you can have a cream sauce … this one always makes us think of peanuts and sesame, duck with peaches … anything with apricots … gorgonzola!'

In the chilly Foreau cellar, the goosebumps come from more than just the cold. Here, one sees the full potential of Chenin's quality, all its varied brilliance and the love that this family feels for its craft. Many of these wines feel as if they will go on forever, not even ageing, merely layering one level of flavour over another.

Domaine François et Julien Pinon

Vernou-sur-Brenne

www.francois-pinon-vouvray.com

To borrow an expression beloved of rural French estate agents, Julien Pinon's winery is *niché dans la verdure*, the cave and modest tasting room barely visible through a mass of trees off a quiet road north of Vouvray. Julien's family has been farming here since 1786 and, like so many French families, raising poultry, pigs and goats alongside cereal crops and vines. Julien is the eighth generation of vignerons. It was only after his grandfather's era that the wine business was separated from the animals and other crops, and Julien's father François took over the vineyards. Under François, who left a career as a child psychologist to run the family business, the domaine went from strength to strength. Like his father, Julien initially worked outside of the wine business, as a town planner, before deciding to return. He trained at the organic

Black flint shines from the limestone in Julien Pinon's cellar

wine school in nearby Amboise and worked at La Grange Tiphaine in Amboise before joining his father full-time in 2016. When François died in 2021 Julien took over the domaine. Viticulture here is organic, certified since 2011.

A visit to the Pinon cellar turns out to be a bit of an Indiana Jones experience. The door opens and a bat whizzes past my face, as if on cue. Through a narrow passageway carved into the rock, dimly-lit walls are studded with fist-sized chunks of black flint that gleam like jewels set into the pale limestone. The prize at the end of the tunnel is, of course, wine, the oldest bottle dating back to 1858. Black flint is a unique part of the terroir of the Vallée de Cousse, where the domaine is situated. These 100-million-year-old soils feature layers of tuffeau and flint, with varying amounts of clay. The climate is cooler in the north of the Vouvray appellation, further from the river and higher up. Seven kilometres and a hundred metres of altitude may not seem significant, but Julien says the grapes ripen a week later here than in the vineyards close to the Loire. This cooler climate has proven to be beneficial in recent warmer vintages, although this later-season benefit comes at the cost of early-season frost risk.

Julien makes a full range of Vouvray styles including a delicious brut *non-dosé* sparkling that spends 4 years on lees, and a *demi-sec* still wine made in solera style using four different vintages. The wines display the coolness of the vineyards. They can verge on lean in lesser years but retain clear and bright acidity in the increasingly common warmer vintages. All are expressive and precise, and hugely pleasurable to drink. Silex Noir is a *demi-sec* wine grown on flint and clay. 'The black flint emphasizes acidity, while the clay gives roundness to the wine,' Julien explains. It is a striking and balanced wine, with mineral-edged acidity and a savoury finish. For me it will always recall the unlikely sight of those jet-black jewels in the cellar. And the bat.

Julien Pinon at his winery

Domaine Huet

Vouvray

www.domainehuet.com

East coast American Sarah Hwang is inescapably an outsider in the small community of Vouvray, but she appears not only to be very much at home, but also to love her vineyards as much as any local. Impending rain notwithstanding, she jumps into her dusty truck to take us to the Clos du Bourg, the much-photographed vineyard that overlooks the distinctive, rounded steeple of Vouvray's church. 'I come here to do yoga and watch the sunset,' she enthuses. It's a pretty special place for a tasting too, and well worth risking the rain to be in the heart of the Vouvray vines.

Sarah arrived to take over the estate in 2012, following the retirement of Noël Pinguet. Her father, Anthony Hwang, bought the domaine in 2003 following the death of Gaston Huet, but his son-in-law, the visionary Pinguet, in charge since 1976, continued to manage the domaine. He converted the 30 hectares to biodynamic viticulture, being an early adopter of the now fashionable idea that this was the best way to maximize the expression of the excellent sites in the family's possession. Although Pinguet's departure in 2012 was a few years earlier than expected, the succession planning was well advanced with an apparently seamless transition to Jean-Bernard Berthomé and now, Benjamin Joliveau, heading up the winemaking. Sarah is keen to stress: 'We are a team here.'

The change of ownership from the Huet family to the Hwang family has not changed the philosophy of either vineyard management or cellar practices. It's all about quality, and yields are low, at an average of 30 hectolitres per hectare. When I ask how they maintain quality, she summarizes: 'We're all pretty OCD here.' For the wines, the aim is to make the best of what the vintage provides. There is no attempt to force grapes into a style that doesn't suit the year, so where weather permits, they pick as late as they need to – by hand, naturally. Sarah says, 'If in 2020, for example, we'd pushed for *sec* wine, something was going to be out of balance – and it's the balance that gives these wines their ageability.' There was no dry wine produced at all in 2020, which she declares was 'a *demi-sec* vintage'.

Sarah embraces the heritage of Vouvray as an appellation of sweeter wines: '*Demi-sec* and *moelleux* are the heart and soul of what Vouvray represents,' she says, 'It's what you can't get elsewhere.' The wines are an admirably consistent representation of the best of Vouvray, with

Sarah Hwang in her beloved Clos du Bourg vineyard

pristine fruit that is generous and ripe but never overblown, and beautiful natural balance at all levels of sweetness. The differences between the terroirs are clear to see, unmasked by oak or other winemaking interferences. There are three *lieux-dits* in the Huet portfolio. Le Haut-Lieu was the first vineyard, bought in 1928 when the domaine was founded. It consists of 15 hectares of deep clay soils on the plateau. Le Clos du Bourg is a 6-hectare plot on clay-limestone soil. The third vineyard is Le Mont, a cooler site with flinty soils that always gives wines with higher levels of acidity. The Haut-Lieu wines have linear acidity, restrained elegance and wonderful purity of fruit. Clos du Bourg offers wines of complexity and richness, powered by mineral acidity. Le Mont takes time to reveal itself, closed and firm in youth, with the brightest of acidity. Sarah says: 'Gaston Huet understood terroir before anyone.' It's a legacy she clearly intends to preserve.

Domaine d'Orfeuilles

Reugny
www.orfeuilles.com

Arnaud Hérivault is the kind of person who looks on the bright side – or so it seems. After battling frost, hail and mildew, he is smiling and full of energy. It's hard to keep up with him as he bounds up the several flights of stairs leading from his winery up into the vineyard, which he calls his 'little corner of paradise'. Arnaud has worked here for 20 years – practically all of his working life, in fact – and lives on-site, as do his parents, as did their parents before them. And it appears to suit him well.

Domaine d'Orfeuilles is not an old property: it was founded in 1947 by Arnaud's grandfather on the site of the former medieval Château d'Orfeuilles. The family has 22 hectares of vines in Chançay and Reugny, many of which are concentrated around the winery. The Loire is 10 kilometres from the vines, which are planted on slopes that are steeper than much of the appellation, with varying exposures: south, south-east and south-west. The river has less of a moderating influence on the climate here, and the presence of frost fans attests to the problems they face. As well as frost, hail is a problem. 'We had hail on 3 June last year,' says Arnaud and I wonder, not for the first time, if winemakers remember their children's birthdays as easily as they recall the dates of significant vineyard events.

The domaine has been certified organic since 2010 and biodynamic since 2020, Arnaud claiming to have seen the beneficial effects of biodynamic practice from year one: 'It's impressive,' he says, 'something you feel.'

As the winery is below the level of the vineyard, making use of gravity is easier than in other parts of the world and the Domaine d'Orfeuilles grapes travel what seems to be a very well-considered journey through the neat cellars, all constructed on a human scale. The wines are well-judged and full of life, retaining mineral freshness even in warmer vintages. The traditional method sparkling, Atemporelle Brut, is a terrific and great-value wine that spends four years on lees. For just a few euros more, the *non-dosé* Feuilles d'Or is also a bargain for a wine that has spent eight years ageing on lees before release. The dry, still wine, Silex d'Orfeuilles, comes from very stony soils. The grapes go through lengthy fermentation, partial oak ageing and lees stirring; the result is mouth-filling, generous and memorable. *Demi-sec* wines display elegance, purity and balance and, even in the sweetest wines, the Chenin acidity remains laser-bright. All feel natural and unforced. Arnaud says he is 'very optimistic' about the future of Vouvray: 'We can do things that others can't. We have a great grape variety that reveals terroir and people.'

Domaine Sébastien Brunet

Chançay

www.sebastienbrunet.fr

Sébastien Brunet explains his approach to winemaking thus: 'I like dry wines, *vins de gastronomie.* When I eat, I like there to be something happening on the plate; with wine, it's the same. I want acidity, bitter flavours, something to wake up my tastebuds. I'm not interested in making wines that are flat.' When one tastes Brunet's wines, this all makes perfect sense. They are anything but flat – and they certainly wake up the taste buds.

Brunet farms 17 hectares of vines in the eastern part of the Vouvray appellation, in the villages of Reugny, Chançay and Vernou. He began to work on a few hectares with his father in 2003 and took over in 2006. Since then he has built up the domaine-bottled wines (much of the production was formerly sold to négociants), bought back vines from

Room without a view: Sébastien Brunet's atmospheric tasting room

family members and converted everything to organic viticulture. Soils are ploughed mechanically and the grapes are harvested by hand. In the cellar, he uses indigenous yeasts, ferments and matures his wines in 400- to 500-litre casks, and typically leaves the wines to mature on gross lees for eight months. The wines have intriguing texture and salinity, bitterness and freshness, richness and finesse. Some of the vines are more than 80 years old and create wines that can send a shiver down your spine. Fosse Rouge is a wine from red clay soils: mineral, with sweet and sour flavours, it is full of personality. Les Pentes de la Folie is like a meal in itself: sweet and salty, full but with acid tension. Brunet is a jovial sort with the air of a man who enjoys life to the full and doesn't take himself too seriously; his wines, though, are seriously good.

Domaine Vincent Carême

Vernou-sur-Brenne

www.vincentcareme.fr

Vincent Carême could be considered a self-made vigneron, establishing himself as one of the stars of the Vouvray appellation since he returned to his native home in 1999. His parents passed on 5 hectares of family vines which they farmed alongside other crops, such as asparagus, strawberries and potatoes, but his was not a winemaking family. Vincent studied viticulture and oenology, and travelled widely to learn his craft. He gained broad experience in France (in Sancerre, Muscadet, Champagne and Alsace) and in the 'other' Chenin Blanc Mecca, South Africa, before returning home to make his mark.

Starting with his 5 hectares, he gradually built the domaine up to 24 hectares today. 'It's the maximum,' he says, although he does vinify other grapes in a négociant operation he runs with his business partner, ex-wife Tania Carême. In 2004 he bought and restored a house in Vernou-sur-Brenne, complete with cellars hewn into the rock for winemaking and storage. Overhead on top of the rock grow the vines of the 2.3-hectare vineyard he acquired with the property and that is the source of his top wine, Le Clos.

Visiting 70-year-old vines in his Peu Morier vineyard, Vincent points out newer vines propagated by *marcottage*, or layering, by burying a cane from an existing 'mother' plant to create a new young plant. 'It's quite common here,' says Vincent, 'and it works well for replacing old vines. He says they don't have problems with phylloxera. 'A bigger fear is that you will plough up the new vine by mistake,' he says. A proponent

Marcottage (layering) in Vincent Carême's vineyard

of organic viticulture, Vincent also leaves grass between the rows of his vines in winter, ploughing superficially in the spring to avoid the grass providing hydric stress to the vines. The soils are shallow, which means greater water stress in times of drought.

Around half of the production is sparkling. Brut Tradition is made by arresting the fermentation when the wine is *demi-sec*. The wine is filtered and bottled with yeast added to provoke a second fermentation. This is essentially a traditional method fermentation except that the sugar that would normally be added is omitted in favour of using the natural residual sweetness from the primary fermentation. The wine spends 18 months on lees before bottling with no dosage. This has lovely lees richness balanced by deliciously crisp acidity. Plaisir Ancestral is a wine produced in the so-called *méthode ancestrale*, a single fermentation that begins in tank and is bottled towards the end of the fermentation, when there are around 18 grams of sugar left to ferment which, Vincent informs me, will give a wine with around 4.5 bars of pressure (compared to the traditional method, where the figure is closer to 6 bars). In this case, no yeasts are added and the wine stays for 12 months on lees. 'It keeps its fermentation aromas this way,' says Vincent. It is indeed expressive and yeasty, with savoury, salty notes – a really tasty wine. The range of still wines includes Sec, a blend of different plots. It is lifted and light, mineral and pure, with a silky, creamy texture and is completely dry. Clos de la Roche comes from a small parcel of vines that Vincent planted in 2012. It is aromatic, mineral and precise, with flavours of honey, lime juice and slate. With higher pHs of recent years, the wines generally all go through malolactic fermentation now and this is no exception. It gives a supple drinkability to the wine without losing any of its focused fruit. Le Peu Morier comes from 70-year-old vines that grow on flinty clay. It has a mineral restraint on the nose. Only in the mouth does it begin to reveal itself, with savoury and salty concentration. With wonderful balance and length, this is super-stylish, modern Vouvray that is so easy to love. Le Clos comes from more calcareous soil. Also quite closed on the nose when young, this is slower to unfurl. Vincent explains that the grapes here are smaller, which gives some tannin to the wine. It is fine-textured and intensely mineral, and needs time in bottle to show its quality.

Vincent Carême is one of the leading winemakers in Vouvray today, and is actively involved with teaching and encouraging young vignerons. He also works with the technical committee of the growers' association on matters such as the best choice of rootstocks, using cover crops and identifying the best *lieux-dits* of the appellation. Never idle, when not tending his vines in Vouvray, Vincent also has a winemaking project called Terre Brûlée in Swartland, South Africa. Here he produces an alternative style of Chenin more akin to those of Anjou noir, grown on granite and schist.

Other Vouvray producers to try

Domaine des Aubuisières, Michel Autran, Jean-Marc Gilet, Vignoble Brisebarre, Domaine Alain Robert (especially sparkling), Domaine Champalou, Clos Thierrière, Domaine des Cormiers Roux, Domaine Florian Le Capitaine, Domaine des Lauriers, Domaine de la Canopée, Florent Cosme, Mathieu Cosme.

Montlouis-sur-Loire

If Montlouis used to live in the shadow of Vouvray, this is no longer the case, at least as far as the French are concerned. Its wines may be less well-known outside their homeland, but the vignerons of Montlouis are doing their best to change this and to highlight the quality of their terroirs, their wines and their commitment to the land.

Just across the Loire from Vouvray, Montlouis-sur-Loire is a much smaller appellation making wines in a very similar style that are 100 per

The Montlouis vignerons created their map in the style of an old school poster

cent Chenin Blanc. The vineyards are sandwiched between the river Loire to the north and the Cher to the south and cover the three communes of Montlouis-sur-Loire, Lussault-sur-Loire and Saint-Martin-Le-Beau. The highest point of the appellation is close to the Loire, so the vineyards enjoy the same gentle slopes and southerly exposure as those of Vouvray and, although the upper soils are different, the tuffeau limestone subsoil is the same. The climate is also similar, although the vines are more susceptible to spring frosts here because the soils warm up more quickly and the subsequent earlier growth is at risk from cold air currents travelling up the Cher towards its confluence with the Loire at the city of Tours.

Montlouis-sur-Loire – key facts

- **AOC granted**: 1938
- **Hectares in production**: 430
- **Annual production volume**: 18,000 hectolitres/2.4 million bottles
- **Styles**: white; 67 per cent sparkling (traditional method and *pétillant originel*), 33 per cent still, at all levels of sweetness: *sec* – less than 5 grams per litre residual sugar; *demi-sec* – between 5 and 18 grams per litre of residual sugar; *moelleux* – between 18 and 35 grams per litre of residual sugar
- **Grapes**: Chenin Blanc
- **Soils**: flinty clay, wind-blown sand, silty sand and silty clay over limestone

An original *pétillant*

One of Montlouis' claims to fame is that it now has its own appellation for sparkling wine made in the traditional way of a single fermentation, which starts in tank and is then transferred to bottle while the wine is still fermenting in order to capture the carbon dioxide released during the process in the bottle. Known as *pétillant naturel* or *méthode ancestrale* elsewhere, the Montlouis Pétillant Originel has been a separate, codified part of the Montlouis appellation regulations since 2007. It contains additional specific quality constraints compared to these other styles. These wines must be made from hand-harvested grapes and, unlike most sparkling wines, which allow for higher yields, those for Pétillant Originel can be no higher than for the still wines. No chaptalization is allowed. The bubbling ferment is then chilled part-way through (to arrest it) and bottled with no added yeast or sugar. The fermentation finishes in the bottle, where the wine must rest for at least nine months on its lees. After disgorgement it must be topped up with the same wine, again with no added sugar and the maximum remaining sweetness (which would be unfermented natural grape sugar) is 5 grams per litre. The wines all require a vintage on the label. The pressure in these single-fermentation wines is lower than for traditional method sparkling, but this is a style also available within the Montlouis appellation.

Less than a quarter the size of its more famous neighbour, Montlouis has nevertheless experienced a resurgence of interest in its wines in recent years, largely thanks to two trailblazing vignerons, François Chidaine and the late Jacky Blot, of Domaine de la Taille aux Loups. As well as making increasingly attention-worthy wines in the Montlouis appellation, the pair also enjoyed significant acclaim for their Vouvray wines. In fact, it is widely thought that this success motivated a successful appeal from the local vignerons of Vouvray to the INAO to ban any Montlouis producer from making wine under the Vouvray appellation unless the grapes were vinified in Vouvray itself. At a stroke, in 2014 such wines were demoted to Vins de France, although the move does not appear to have done any particular damage to the reputation of, or demand for, these wines. There may be some logic to the INAO decision, but it is inconsistent with other appellations. In Savennières, for example, there are multiple producers whose cellars are across the river in the villages of the Layon Valley (Domaines des Baumard, FL, Belargus and Ogereau, for example). They have parcels of grapes in

Savennières and others in other parts of Anjou, and vinify everything centrally at the cellars in their home villages.

There is a strong organic culture among the vignerons of Montlouis, where 65 per cent of the 48 growers are certified organic (compared to Vouvray, where the number is closer to 20 per cent).

Notable Montlouis producers

Domaine de la Taille aux Loups

Montlouis-sur-Loire

Jacky Blot was a visionary winemaker and a leading figure in changing perceptions of Montlouis wine. A former wine broker, he came to Montlouis for two reasons: first, because it was affordable, and second, because he believed he could make great wine here. This was in 1988, and he started the Domaine de la Taille aux Loups with 5 hectares of vines. His focus – unusual at the time – was to make dry wines from single vineyards. In the 1990s it was not fashionable to produce totally dry wine here, so his approach marked him out from the start. Everything was harvested by hand and sorted meticulously. Jacky understood the importance of careful preparation and attention to detail: in a former life he had been a professional parachutist.

Following Jacky's death in 2023 his son Jean-Philippe, who had been working with his father for a number of years, took over the domaine, which now comprises 45 hectares. It includes two walled clos in Montlouis (Clos Michet and Clos de Mosny) and 5 hectares in Vouvray, Clos de la Bretonnière and Clos de Venise. The latter wines have been classified as Vin de France from the 2014 vintage, when the INAO ruled that any wine made from grapes grown in Vouvray needed to be vinified in Vouvray too (see above).

The Domaine de la Taille aux Loups wines are precise, rich and mostly dry, expressing terroir rather than winemaking. Yields are kept low, and at harvest, the grape sorting starts in the vineyard. After pressing, the juice is settled twice before fermentation in Burgundian barrels in the cool stone cellars. The work is done by gravity, with the reception of grapes on the third floor and the press below on the second floor. The fermentation and ageing are done with no additions of sulphites, ensuring that the barrels are kept topped up to avoid oxidation.

The domaine produces a wide range of wines from the terroirs of Montlouis and Vouvray. Remus is the entry-level wine, a blend of small lots around Montlouis. With a baked apple character, it is quite soft and delicate, with a smoky edge. Clos Michet comes from a sunny spot and is a very approachable wine. 'It's for people who don't know Chenin,' says Jean-Philippe. Clos de Mosny, from a 12-hectare south-facing site of sandy clay and flint over limestone, has a delicious richness, with yellow peach and nectarine fruit. Les Hauts de Husseau is from a 6-hectare site close to the village of Montlouis, at the highest point of the appellation. Here the soils are thin, and the wine has fine mineral intensity and length. The two Vin de France wines from Vouvray are notable, both coming from the privileged *première côte*. Clos de la Bretonnière is a south-east-facing 4-hectare plot on thin clay soils over limestone. An upright wine, it has lovely depth of flavour and zingy minerality. The Clos de Venise is a full south-facing site that ripens a week earlier than the Clos de la Bretonnière. It is rounder, peachier and altogether softer in style, with honeyed appeal.

The domaine is probably as well known for its sparkling as for its still wines. In particular, Triple Zéro was a groundbreaking wine when it first appeared in 1993, although at the time, it didn't go down too well. This was perhaps because the wine was completely dry, and at the time the fashion was for sweeter fizz. And 1993 was not the ripest of vintages, which probably didn't help. The Triple Zéro is a single-fermentation *pétillant* wine whose name comes from the fact that there is no sugar added at any of the three stages where it could be. The grapes are picked at a high level of ripeness and there is no chaptalization when they first start to ferment, no *liqueur de tirage* to continue the fermentation in bottle, and no dosage after the wine has been disgorged. This way of making sparkling wine was subsequently codified as part of the Montlouis appellation as Pétillant Originel, in 2007 (see above).

François Chidaine

Montlouis-sur-Loire

www.francois-chidaine.com

François Chidaine's father was a Montlouis vigneron, tending 4 hectares of vines locally. He passed the business on to his son François in 1989. But François had bigger – and better – ideas of what he wanted to achieve. This small local winegrower has gone on to become the most famous son of Montlouis and one of the most respected vignerons in the Loire Valley, producing wines from 45 hectares in Montlouis, Vouvray and the wider Touraine area.

François wasted no time in converting the domaine to organic viticulture, and by 1999 the vineyards were also certified biodynamic. He

Montlouis trailblazer François Chidaine

grew his landholdings in Montlouis to 22 hectares, spread over 30 different plots and, turning his attention north of the river, managed to acquire 8 hectares of prime land in Vouvray. His aim from the outset was to craft wines that show the nuances of each individual terroir, so the number of cuvées quickly multiplied, and today he makes more than a dozen wines across the two appellations. From the 2014 vintage, the wines grown on Vouvray plots (notably Les Argiles and Baudoin) are labelled Vin de France – see page 173 for the reason for this.

François remarks, 'In the last 30 years, viticulture has changed to become more professional, so we need less oenology and shorter élevages.' Most of the work is done in the vineyard. The hand-harvested grapes are sorted here, sometimes several times, and brought to the cellar in small crates. After a long and gentle pressing, they ferment slowly with indigenous yeasts and are aged on the lees in large-format oak, all in the naturally chilly tuffeau cellars.

François Chidaine's sweetness scale appears on all of his bottles

The domaine produces a range of styles from fully dry to sweet and sparkling. Most helpfully, from a consumer's perspective, each back label shows an '*échelle de sucrosité*', indicating the level of sweetness that one can expect from each of the wines in the range.

The wines are elegant and beautiful expressions of Chenin Blanc. Those from Montlouis show the more delicate and subtle side of Chenin. The Vouvray wines are more open in style, though no less elegant, and there is a juicy fruit purity evident throughout the range. Personal favourites include the outstanding Vouvrays (the mineral, tense and chalky-textured Les Argiles and the powerful, elegant Baudoin) and from Montlouis the floral and airy Clos Habert, a '*tendre*' style from flint-laden soils, and Les Bournais, a dry and delicate wine with beautiful balance.

Le Rocher des Violettes

Dierre

www.lerocherdesviolettes.com

Le Rocher des Violettes is a family affair. Xavier Weisskopf runs the winery, his wife Clémence runs the office, their young daughters run the office décor (paintings everywhere) and their little dog – Sulphite – runs about like a mad thing (I later learn she is in serious training to become a truffle dog). Life could have been quite different for Xavier: he started out studying agrochemicals and crop protection, but had a change of heart and decided that winemaking should be his *métier*. And while he was at it, he ditched the agrochemicals altogether.

After studying in Beaune Xavier, young and keen, but with only a little experience under his belt, managed to acquire 7 hectares of Chenin Blanc in Montlouis along with a parcel of centenarian Côt vines. This was in 2005. By 2008 the domaine was certified organic and today has grown to 17 hectares, which includes 14 hectares of Chenin as well as some Chardonnay, Cabernet Franc and Pinot Noir.

The Rocher des Violettes winery is in Dierre, 3 kilometres to the east of Saint-Martin-Le-Beau, where the vines are located. This is the far east of the Montlouis appellation. 'It's more frost-prone, and therefore cheaper to buy land here,' says Xavier, adding, 'I didn't know that when I bought it.' The vines grow close to the Cher river on varying types of sandy, silty clay and the locally named '*perruches*' soils which are a mixture of sandy clay and flint, all over limestone. The vine's growth starts early here, so Xavier is ready every year with candles and piles of wood to mitigate the effects of possible frost damage to the young shoots. There is also a frost fan looming overhead to protect a part of the vineyard. Another strategy employed to protect the young shoots is double pruning (a lighter pruning before potential frost and a final pruning when frost risk has passed), which is effective, but costs, he says, 25 per cent more to implement. Further up the slope, away from the river, the soils are more alkaline, 'It's perfect for truffles,' he says. Sulphite had better take her training seriously.

In the vineyard, Xavier has been experimenting with different types of vine training in an attempt to reduce the susceptibility of younger vines to the trunk disease, esca. But he has his own theory as to why some vines are more vulnerable than others, and it's all down to how the vine is pruned, or more specifically who is pruning it: 'I have a theory it's due to the excessive change in personnel pruning the vine,' he says, noting that he has seen a 40 per cent increase in susceptibility to the malady in plots where there has been change in the people doing the pruning.

The still wines are site-specific cuvées, all harvested by hand. The juicy Touche-Mitaine grows on grey clay soils and spends a year in old oak. Xavier is not a fan of new oak for Chenin: 'It can't take it; Chenin is very easily marked by oak,' he says. La Rochère, very close to the Cher, is a south-facing slope of silty clay over limestone originating from lake sediment from the Ice Age. These are white soils that give a lot of mineral tension. The wine is fermented in a concrete egg and shows a rounded core of aromatic fruit balanced by electrifying acidity. Xavier recommends waiting two years before approaching it, when the acidity will soften, but anyone who enjoys the thrill of firm, stony acidity like this will love it from the start. Le Grand Clos comes from deep, stony, red clay soils, and is a late-ripening site. The wine is fermented and aged for a year on its lees in 500-litre barrels, where it is stirred periodically. This gives a wine that has body but remains firm, with delicious texture and a floral character. Les Borderies is a *demi-sec* style with around 15 grams per litre of residual sugar. Grown on sandy soils, it recalls a Mosel-style Riesling, with its light, floral, effortless elegance and juicy deliciousness. Another sweet style, a *moelleux*, is Le Grand Poirier. Made with slightly dried (*passerillé*) grapes that grow on soils that comprise red clay, wind-blown sand and silt, it has a gentle character and light, honeyed sweetness.

Xavier also makes one of the best Montlouis Pétillant Originel wines, with elegant, fine bubbles and clear fruit. The domaine's red wines come from the same sites as the whites, from vines planted as long ago as 1891. The tiny 0.1-hectare of Côt, which includes the ancient vines, produces an aromatic, black-fruited, lightly structured and fresh wine, while the Cabernet Franc needs more work to soften the tannins, and has at least two years' ageing. Less exciting than the whites, perhaps, they nevertheless show the characteristic elegance of the house style. Xavier also makes an orange wine from Chenin, Pinot Noir and Chardonnay, and a Côt rosé.

Les Pierres Ecrites

Saint-Martin-Le-Beau

Anthony and Coralie Rassin are relative newcomers to Montlouis and this is the first winery they have owned, but they are no strangers to winemaking. Both had a decade's experience working in the Rhône for Jaboulet Aîné (Coralie) and Cave de Clairmont (Anthony) before deciding, in Anthony's words, 'to write our own story'. They would have stayed in the Rhône, he admits, 'but there was the problem of zeros'. Loire vineyards were more attractively priced. The pair took over the domaine Flamand-Delétang with 8 hectares of vines in the eastern part of the appellation in 2015, changing the name to Les Pierres Ecrites.

The story does not have an auspicious start. 'Our first year, 2016, we lost 95 per cent of the grapes to frost … and then again in 2021 – but we are still smiling,' says Anthony. 'As far as the climate goes, we are in the right place – and we can react.' Now, they prune twice – before and after frost – to try to protect the vines, although Anthony would prefer not to have to do this: 'It weakens the vine,' he says, 'because it bleeds its sap. You are taking the vine's energy reserves when you do this.' Despite the challenges of frost, Saint-Martin-Le-Beau also has its advantages. 'The deep soils give us aromatic wines,' he says, 'and they are good at protecting the vines in times of drought.'

They started converting the domaine to organic, and then biodynamic viticulture. They work as a team and do all the vineyard work by hand. Wines ferment in 400-litre barrels and they use no sulphur until bottling. Their dry white, Empreintes, grows on a vein of clay. 'After fermentation we don't touch it for a year,' Anthony says. The wine has super fruit concentration and manages to be both delicate and forcefully mineral, with real substance. It feels light in the mouth but has huge persistence on the finish. Chenin du Berger is a *demi-sec* wine made from the fruit of 80-year-old vines that grow on deep sand with a little flint. It is lovely: soft, floral, easy and effortlessly light with honey and pear fruit. Or des Petits Boulay is a *moelleux* wine made with grapes that have dried on the vine. The location is not favourable for botrytis, as they do not get the required autumn mists that promote its development. This is the same terroir as for Empreintes, and the fruit shows a similar pear (and dried pear) character. The earlier, ripe 2018 vintage of this wine is a stunner. It has apricot and ripe apple flavours, edging even towards pineapple, a touch of vanilla from the oak, and all this concentrated flavour with the hallmark weightlessness of fine Montlouis.

Les Pierres Ecrites is a domaine that is rapidly joining the ranks of the best in Montlouis. With two winemakers at the helm, I wonder if the decision-making is always a harmonious process? 'We always agree, yes,' Anthony replies. 'She wanted amphorae, I didn't. So we got them, and she was right.' It sounds like a successful partnership.

Other Montlouis producers to try

La Grange Tiphaine (see p. 235 for more details), Laura David, Bertrand et Lise Jousset.

When in Vouvray and Montlouis

Places to stay

Hôtel les Hautes Roches, Rochecorbon (www.leshautesroches.com). Just a couple of kilometres along the road from Vouvray and right on the river, this hotel combines elegant eighteenth-century architecture with rooms built into the rock for a luxury troglodyte experience. The Michelin-starred restaurant serves classic and imaginative French dishes that are naturally designed to pair with the local wines. The shady balustraded terrace overlooking the river encourages lingering lunches. It's hard to imagine a better place to enjoy soaking up the atmosphere of Vouvray while sipping a glass.

Le Gaimont chambres d'hôtes (www.legaimont.fr). Hidden behind high walls and tall gates close to the centre of Vouvray, this is a super place to stay. It's also another chance to sleep in a cave or, even better as far as I'm concerned, to bathe in one. As well as the caves there are very comfortable rooms in the main house. If you can drag yourself into the daylight from your cave cocoon, there is a garden to die for and an outdoor pool heated to comfortable body temperature to encourage lounging. The place is a delight even before you meet the genial and attentive owner, Matthieu. He and his charming wife Elisa have thought of every comfort and and everything is done with style. Candles and jazz light up the breakfast room. Staying here you feel like a special guest in someone's much-loved home.

Places to eat and drink

All the best things in Vouvray are found in caves, and here's another: Les Gueules Noires (gueulenoirevouvray.wixsite.com/les-gueules-noires) is a Michelin-starred troglodytic restaurant in the heart of the Vouvray vines that is the most-demanded place to eat locally (and with limited opening hours, it is advisable to book well ahead). Very informal, it has a fabulous wine list and a garden where meals are served in summer under the shade of the linden trees. In nearby Rochecorbon, Café Corbon is a casual but stylish bistro for a light lunch of sharing plates and *plats du jour*. When in Montlouis, a good bet is Le Berlot in the main square. It has creative and more traditional dishes and a wide-ranging wine list.

For a picnic or edible souvenir of Vouvray, the red neon signage of Charcutier Hardouin is impossible to miss, by the traffic lights at the entrance to the village. As well as all manner of meat products, they sell a range of coarse rillette patés (a Tours speciality), tarts, quiches, salads and other prepared dishes – and even a few bottles of Vouvray. For a wider choice of wines (and whisky, cognac and rum), and local produce that includes cheese, fruit, vegetables and all manner of delicious things in jars, Au Jardin Vouvrillon (www.jardinvouvrillon.fr) might just be a contender for the best corner shop in the world. These are especially useful options when visiting Vouvray at the start of the week, when almost all of the restaurants are closed.

To taste – and buy – the wines of François Chidaine and a few other selected vignerons, plus a selection of delicatessen items, head to the Montlouis Loire quayside to La Cave Insolite, run by François's wife, Manuela (www.francois-chidaine.com/la-cave-insolite).

CHINON, BOURGUEIL AND SAINT-NICOLAS-DE-BOURGUEIL

Chinon

Spend any time in Chinon and it won't be long before Rabelais crops up. The author, physician and most famous son of the city is also its greatest source of pride. Chinon and Rabelais are simply inseparable in the eyes of the locals – even if the wines may have changed a little in the 500 years since he was drinking them and praising them in his fictional tales. In 2016 the winemakers even managed to get the boundaries of the appellation redrawn and expanded to include the full area of

Chinon vines overlooking the castle and the Vienne

the imaginary land of Rabelaisie where the author based the adventures of Gargantua and Pantagruel. When Rabelais penned his epic works, could he ever have imagined that he would become the major marketing man for the winegrowers of Chinon in the twenty-first century? Critics might say that today's Chinon producers are more focused on the past than on the quality of their current wines. But when Chinon wine is good, well, Rabelais was right: it is worth writing about.

The largest red wine appellation of the Loire Valley spreads around the town of Chinon, which is situated on the banks of the Vienne. This is one of the Loire's major tributaries, which flows from the south-east. The heart of the vineyard area lies in an elongated triangle between the Loire to the north and the Vienne to the south. Fewer vineyards lie to the south of the Vienne. At the westerly tip of the appellation, the rivers merge, at Candes-Saint-Martin. The gently undulating slopes carved by these large bodies of water range from 35 to 110 metres in altitude. The climate is oceanic to the west, becoming semi-continental towards the east of the appellation, and is moderated both by the influence of the two rivers and by the substantial forest of Chinon immediately north and east of the vineyards, which protects the vines from cold winds from the north. The appellation enjoys the warmest and driest climate in the Touraine region. Summers are sunny with little rain, making this a natural choice for the red wines that make up the vast majority of Chinon's production.

Chinon – key facts

- **AOC granted**: 1937, enlarged 2016
- **Hectares in production**: 2,354
- **Annual production volume**: 90,000 hectolitres/12 million bottles
- **Styles**: 86 per cent red, 9 per cent rosé, 5 per cent white
- **Grapes**: Cabernet Franc, Chenin Blanc. Cabernet Sauvignon is permitted up to 10 per cent but most reds are 100 per cent Cabernet Franc.
- **Soils**: three main types – alluvial terraces of sand and gravel close to the river; slopes of Turonian yellow limestone and clay; higher plateaus of siliceous clay, silt and sand over Senonian limestone.

Chinon wines can be among the most structured and long-lived of the Loire, although it is perhaps their succulence that ranks highest in their appeal. Good producers are reining back on oak and tannin extraction to produce highly attractive wines that don't demand patience to be enjoyed. Terroir variations account for differences in styles too, of course. Low-lying sandy alluvial terraces close to the Vienne give light, aromatic and floral wines for early consumption, while those from the limestone slopes are characterized more by dense, black fruit and structured tannins. The subsoil, as throughout the Touraine region, is limestone; it runs a hundred metres deep here. But in many places, the soils at the surface are visibly different from the pale, chalky vineyards of the Saumurois. Chinon soils are often yellower and warmer, producing richer, more generous red wines, which are typically rounder and less austere than the Cabernet Francs of the Saumurois can be.

Cabernet Franc rules supreme in Chinon, where it dominates the viticultural landscape. Locally known as Breton, the origins of this name are unclear, although multiple theories have been put forward (unsurprisingly, in Chinon it is attributed to Rabelais). The rare whites are from Chenin Blanc, a mere 5 per cent of vines at the time of writing, although some producers declare an interest in increasing their plantings of the grape. The relatively warm climate produces a riper style of white wine with more measured acidity than is found elsewhere in the Loire Valley. These easy-drinking wines, always dry, are found locally, but quantities are small and prices tend to be high.

Loire Cabernet Franc: finding its identity

The Loire's number one black grape is an ancient variety, most likely the oldest in France today. Its origins are uncertain, but it is thought to have originated in the Basque country in northern Spain, from where it travelled to Bordeaux and on to the Loire. There is still plenty of Cabernet Franc in Bordeaux, of course, but it almost always plays a supporting role to its progeny, Cabernet Sauvignon, or to Merlot. It ripens earlier (and therefore more reliably) than Cabernet Sauvignon and is valued for its ability to add a floral perfume to a blend.

In the Loire, Cabernet Franc was long since adopted as the red wine variety of choice for the terroirs of Touraine and Anjou. Red wines from Chinon,

Bourgueil, Saint-Nicolas-de-Bourgueil and the Saumur appellations are usually 100 per cent Cabernet Franc.

The variety has undergone something of a renaissance in the last few decades. In the early 2000s, Master of Wine and consultant Sam Harrop worked with Cabernet Franc growers to explore different options for vinifying the Loire's signature black grape. 'I wasn't trying to tell them what to do, but I was trying to get them to find their own identity,' he says, adding, 'There was a tradition of picking the fruit before true ripeness and extracting the hell out of it.' After decades of routinely excessive fertilizer use in the 1960s and 1970s, and higher permitted yields than today, the resulting wines were often hard to drink: light on fruit and heavy on tannins, with green, leafy flavours, especially in cooler years when it was challenging to ripen the crop. You might say it was a case of style over terroir, as many growers, seeing the success of Bordeaux reds, tried to emulate a bigger, oakier style. 'I encouraged them to look more to Burgundy than to Bordeaux,' Sam says.

Jacques Dutheil (of Couly-Dutheil in Chinon) was interested in experimenting with a new approach. He worked with Sam to produce a trial of wines made from grapes harvested on the same date that then underwent different winemaking interventions. They were able to demonstrate the effects of cooler fermentations, lower levels of extraction, and of using or not using oak for ageing. 'The idea,' explains Sam, 'was to celebrate the acidity rather than the tannins.'

There is now a greater focus on terroir expression throughout the Loire, and this includes Cabernet Franc styles as well. While there is still a tendency in some domaines to make more structured and extracted wines with extended oak ageing, more often wine growers are realizing that this is not always playing to their natural strengths. The Loire remains a marginal climate for red wines and will never be able to compete with more full-bodied, higher alcohol, richer reds from the south. And it has no need to. What the Loire has to offer is a rarer commodity: wines that guarantee refreshment, the acidity Sam Harrop was talking about. It comes from the cooler climate and from the tuffeau limestone soils. It comes from picking the grapes at the perfect moment and then handling them with care, so that the fruit is to the fore.

Cabernet Franc vignerons have benefited from the warmer vintages of recent years: 2018, 2019, 2020 and 2022 gave them grapes with excellent ripeness levels and the wines have never been better. Finally, it looks as though Cabernet Franc has found its own identity in the Loire Valley.

Notable Chinon producers

Château du Petit Thouars

St-Germain-sur-Vienne

www.chateaudptwines.com

Without doubt one of the most beautiful properties in the Chinon appellation, Château du Petit Thouars is tucked away in the newest and most westerly corner of the appellation, close to the Vienne's confluence with the Loire. Set in 150 hectares of trees, gardens and farmland south of the river, the property has been in the du Petit Thouars family since 1636. Family records show there was a vineyard planted on the site in the distant past, but the modern story of the domaine's wine begins with Yves and Marguerite du Petit Thouars who, in 1978, planted the 16-hectare vineyard that adjoins the château today.

The first time I visited here, in the 1990s, it was to be hosted for a memorable dinner by Marguerite, a wine writer and colourful personality who one of our party found so beguiling he had to be restrained from carrying her off up the stairs at the end of the evening. Today it is her son, Sébastien, who has taken over the reins. Another vivacious character, he enjoyed a very different life running an electronic music label in Paris before returning to the family home to help his father with the winery in 2009. Today, he and his Canadian wife, D'Arcy, have made the vast edifice their family home. While the pair have embraced the country château lifestyle with their three young children (and let's face it: who wouldn't?), Sébastien has clearly lost none of his passion for music and is almost as keen to discuss his record collection as his wines. His definition of the character of Chinon wine is one that could equally be applied to himself: 'vibrancy, generosity and the Rabelaisian spirit of enjoying life'.

Typical of the appellation, production here is almost all red from Cabernet Franc, with a few hectares of Chenin for whites. A small amount of sparkling and rosé is also produced. There are three main red cuvées, distinguished by the age of the vines and winemaking style, rather than any particular difference in terroir. Les Georges is a light, early-drinking style made purely from free-run juice. Aged for a short time in stainless steel tanks, it offers primary fruit expression. L'Epée contains some press wine and spends some time in aged oak to give more structure and longevity. The top cuvée of the estate is L'Amiral. Made from pure press wine and produced only in good vintages, this is a wine with strong oak influence and is intended for long ageing. The white Les

Château du Petit Thouars: when your home is your castle

Clos is barrel-fermented and lees-aged. Made from young vine fruit that is hand-picked, this is a wine that is destined to improve as the vines age.

With a younger generation at the helm, the winds of change have been blowing through this traditional estate. Recent developments include a transition to organic viticulture starting in 2019 and hiring a new young winemaker, Clément Curet, from the 2021 vintage. Curet says he is aiming for gentler extraction in the red wines, subscribing to modern 'infusion' methods that tame the sometimes rustic tannins of Cabernet Franc and allow the fruit to shine. This bright, approachable, juicy style suits this cooler part of the appellation well and the estate appears to be on an upward trajectory.

Clos des Capucins

Chinon

www.closdescapucins.fr

Fiona Beeston might just have the best view in Chinon. From her kitchen, she can survey the sprawling castle of Chinon and below, the Vienne, her favoured swimming spot. The Clos des Capucins is a bijou property of barely a hectare, located on a site where monks of the Capuchin order settled in the early 1600s.

Fiona's long previous career in wine had centred on Paris, where she started out working for crème de la crème wine merchants Steven Spurrier at Les Caves de la Madeleine and then Lucien Legrand who, she says, 'taught her everything'. She later worked as a wine journalist and author. At the weekends, she would often come down to the Loire to buy wine for her employers, and grew to love the area. So when she decided to try her hand at making wine herself, the Loire was a natural choice. She acquired the Clos des Capucins in 2010, following a three-year search for a small vineyard somewhere in the Loire. 'All of the good white terroirs were taken,' she says, 'so I turned to reds.'

She chose Chinon for the 'voluptuousness' of its wines and looked (unsuccessfully) for a vineyard to purchase on the favoured slopes of Cravant-les-Coteaux before turning her attention to 'places where the monks had been.' She was assisted in her endeavours by her friend and winemaker Charles Joguet. 'I came here because of him,' she admits. 'He led me here.'

The *clos* consists of four small parcels, surrounded by hedges and trees, the whole enclosed by a wall. The land includes a small patch of centenarian vines immediately in front of the house. Other vines are 50 to 60 years old, with some newer plantings from the 1980s. Fiona has also bought another hectare or so of vines a couple of kilometres away. These grapes, farmed organically, go into her second wine, Perfectly Drinkable Chinon. This is a fragrant and juicy wine with black cherry fruit and a meaty edge – perfectly drinkable Chinon, indeed.

The 2018 Clos des Capucins Eminence Grise is an intriguing wine. With a smoky nose from ageing in large oak, its fruit is youthfully exuberant four years on, with notes of cherries and fresh flowers. But there is also tension, texture and structure here, a reflection, Fiona suggests, of her own character. 'I'm English – upright,' she says. The name Eminence Grise is a reference to Père Joseph, François Leclerc du Tremblay, who founded the monastery and became the original *éminence grise*, mentor to Cardinal Richelieu. But it is also a reference closer to home: her own right-hand man, Charles Joguet, who continues to consult on winemaking. Production is low and, understandably, quickly snapped up by private customers and restaurants.

Couly-Dutheil

Chinon

www.coulydutheil-chinon.com

The story goes that Couly-Dutheil was founded on the love of a horse. Baptiste Dutheil had bought the creature and was very fond of it. His wife, however, was less impressed and asked him to sell it. Ignoring her request, he instead bought a plot of land opposite the castle in order to have a reason to keep his equine companion: how could he plough his field without a horse? It appears this strategy won her round (there is no talk of further dissent, anyway). The horse remained, vines were planted, wines were made. And the rest is history. From small beginnings in 1921, Couly-Dutheil is now the joint largest producer of Chinon wine and that original vineyard – the Clos de L'Echo – one of its most coveted sites.

Today the domaine is run by Arnaud Dutheil, the great-grandson of Baptiste. Arnaud's late father, Jacques, was instrumental in bringing this historic domaine's style into the modern era in the early 2000s. Working with consultant Sam Harrop MW during the years of the Cabernet Franc project, a collective effort to modernize and improve the Cabernet Francs of the Loire (see box, p. 184), he experimented with less extractive winemaking, an approach Arnaud continues today. There is little wood ageing and filtrations are light. There is an emphasis on using indigenous yeasts and as little sulphur as possible in the winery, all with the aim of allowing the fruit to come to the fore.

Viticulture is not organic but in an effort to maintain biodiversity in the vineyards and reduce erosion, they plant grass between the rows and minimize use of agrochemicals. The philosophy of the domaine is to harvest later than most, ensuring the grapes are fully ripe to avoid the vegetal character often associated with Cabernet Franc wines of the past.

The range of a dozen or more wines reflects the many different sites across the 97 hectares owned by the domaine. These include rosé and sparkling wines, white from Chenin Blanc and even a white wine from Cabernet Franc. Since 2010, whites have been an increasing focus, with 10 per cent of the vineyard now planted to Chenin (more than double the appellation average), but reds are still very much to the fore.

The flagship wines come from the domaine's two *monopole* sites, Clos de l'Olive and Clos de l'Echo. Clos de l'Olive has more limestone, Clos de l'Echo more clay, but both are very good terroirs, giving wines of power, elegance and great longevity.

Situated in the heart of the city, Couly-Dutheil is a worthy ambassador for the appellation and easily accessible to visit year-round.

Domaine Bernard Baudry

Cravant-les-Coteaux

www.bernardbaudry.com

One of the most respected names in the Chinon appellation, Bernard Baudry established his estate in 1975 with just two hectares of vines. The domaine has since expanded to encompass plots amounting to 32 hectares, all farmed organically since 2006. His son Mathieu joined in 2000, after chalking up experience in Burgundy, California and Australia.

All of the main terroirs of Chinon are represented in these wines: sandy gravels in the valley, chalky clay on the slopes and sandy limestone on the plateau, though Mathieu points out that there are variations within each of these. The grapes are harvested by hand, destemmed and fermented using native yeasts. Ageing is in cement vats, concrete eggs or oak barrels, depending on the cuvée, but no new oak is used. Mathieu explains that the high humidity in the cellar reduces the need to top up the barrels and reduces the risk of oxidation during maturation. Most of the wines are bottled unfiltered.

Bernard Baudry wines are full-bodied and savoury, and have attracted a loyal following over the years. The domaine has some excellent terroirs, including the flagship La Croix Boissée, whose wine ages for two years in barrel and a further year in vat. Recent tastings have proved irregular in quality, perhaps due to low levels of sulphur used at bottling leaving the wine more at risk of *Brettanomyces* infection; nevertheless, the domaine remains a reference for the appellation.

Domaine Charles Joguet

Sazilly

www.charlesjoguet.com

Charles Joguet, an artist and poet by vocation, took over this domaine on the death of his father in 1957. With no prior experience, he spent the next forty years building his reputation as an innovative producer of quality Chinon. His name remains synonymous with quality today under the Genet family, who have owned and managed the property since Joguet's retirement in 1997.

Joguet was an early adopter of parcel-by-parcel harvest and vinifications in the Burgundian style, with all grapes harvested by hand. He was also the first to install stainless steel tanks in the region, at the same time recognizing the natural resources available under the ground: the wines age in a magnificent and cavernous fifteenth-century former quarry that offers perfect conditions of constant temperature and humidity. Quality improvements have continued since Joguet's retirement from the business. The Genets have sold lesser plots and bought others, all while keeping the estate to a manageable size. The vineyard has been certified organic since 2013. The property currently stands at 35 hectares, spread across a 25-kilometre area that includes some of the greatest terroirs of Chinon.

The range includes seven reds, two whites and a rosé. Fermentations tend to be cool and there is restraint in the use of oak for ageing. While some of the wines are destined for immediate consumption and others for longer ageing, there is always a distinctive brightness and ease in Charles Joguet wines. Commercial director Anne-Charlotte Genet says, 'The DNA of Loire wines is freshness. We're not looking for powerful, rich wines. We are not the Rhône and there is no point trying to be.'

Silènes, from a 10-hectare plateau between the Loire and the Vienne, is a crisp, frank and open expression of Cabernet Franc that is a perfect entry point to the house style. Petites Roches, from more gravelly soils, has more substance, yet retains a clean-edged precision. Both of these wines are cool-fermented in stainless steel to retain their bright, primary fruit quality.

The more structured cuvées come from grapes grown on soils with more gravel, clay and limestone. La Cure is silkily textured and feels positively alive with freshness, while Les Charmes, from a north-facing hilltop site on the left bank of the Vienne, offers more black cherry intensity, savoury tension and a touch of oak spice, while retaining a delicious juiciness on the finish. To call this a more serious wine somewhat detracts from the fact that it is just plain fun to drink. From the same vineyard comes a delicious white, Les Charmes, a Burgundian-style oak-aged Chenin with chalky, lemon zest fruit and creamy richness.

At the upper end of the Joguet range, Les Varennes du Grand Clos is a wine of understated elegance that comes from old, low-yielding vines south of the Vienne. It combines the fruitiness of a sandy terroir with the structure of clay-limestone. Cold maceration of the grapes and long, cool fermentation ensure maximum fruit expression, while ageing in predominantly old barrels adds complexity. The two flagship cuvées of the domaine come from vineyards that sit on opposite sides of the Vienne river. Clos du Chêne Vert is one of the oldest vineyards of Chinon, on a steep, south-facing slope above the town. It is easy to see why this was one of the first sites chosen to plant Cabernet Franc. The wines are generous, floral and expressive, with plump, ripe fruit. While immediately appealing, the wine is certainly worth waiting for. The second is from the, Clos de la Dioterie Monopole vineyard, which faces north-east. Here, the low-yielding, ancient vines give a stricter and initially more closed style that takes time to unfurl: the intense fruit is dark, savoury and tightly wound.

Domaine Charles Joguet offers a fabulous range of fruit-focused wines that reflect the varied terroirs of Chinon. Ranging from the every-day to the sublime, these are wines of clarity and precision that merit the attention of any wine lover.

Domaine de la Noblaie

Ligré
www.lanoblaie.fr

The sign outside Domaine de la Noblaie advertises 'Le Charme du Chinon', which pretty much sums up the people and the wines of this delightful domaine, with its undulating vineyards on the rural left bank of the Vienne. The Billard family have been farming this land since 1952, although maps show that La Noblaie was identified as a named vineyard site as early as the eighteenth century.

Today the 27-hectare estate is run by genial winemaker Jérôme Billard and his agronomist wife, Elodie Peyrussie. Jérôme is one of a growing number of local winemakers with an impressive international CV that allows him to bring a modern, global perspective to his craft. He studied oenology in Dijon before working with France's 'other' Cabernet Franc on Bordeaux's Right Bank, with the Moueix family, latterly gaining international experience with their California icon wine, Dominus Estate. He also worked at Sacred Hill in New Zealand.

Jérôme explains that the left (south) bank of the Vienne is quite different from the more recent, more intensive viticulture of the right bank: 'Here, we have only small blocks of vines, interspersed with bushes, truffle oaks and crops. Most of it was developed centuries ago.' In addition to vines, the family owns sheep and farms a further 30 hectares of cereal crops.

The organically grown grapes are harvested by hand and transferred by gravity to the cellar. Jérôme avoids harsh extraction of tannins, favouring a gentler approach that reveals the individuality of the terroirs in the wines.

White wines are important here, accounting for 20 per cent of production. Chant Le Vent is a pretty, crisp style made from the fruit of 20-year-old vines. Fermented in stainless steel, it is enriched by a few months spent on lees before bottling. Similar production methods are used for another fresh style, the Goutte de Rosé. This is a 100 per cent Cabernet Franc rosé intended for early consumption and bottled, reassuringly, under screwcap, a rare sight in the Loire Valley. It's fashionably pale, completely dry and really good. The second white, La Part des Anges, comes from 60-year-old vines that grow on the tuffeau slopes of the domaine. It undergoes a long, slow fermentation and maturation, initially in barriques, before ageing for up to 18 months in large oak vats. This is a wine that can age for a decade, gaining in fullness and stature over time, the richness balanced by the incisive freshness that comes from chalk-based soils.

Reds include the juicy, fresh and aptly named Le Temps des Cerises – simply lovely – and more structured, age-worthy cuvées such as the complex and spicy Blancs Manteaux and Pierre de Tuf, the latter from 70-year-old vines that are fermented in an ancient vat dug into the tuffeau. Pierre de Tuf is a powerhouse of flavour, complexity and finesse, with enveloping sooty tannins and a long finish.

The wines of Domaine de la Noblaie are full of personality and interest, reflecting both a modern, quality-driven approach and the rich diversity of the soils. Jérôme describes a soil study of their vineyards that entailed digging 800 holes. 'It showed us we had 20 different types of terroir,' he says, 'a bit like an underground lasagne.'

Domaine de Pallus

Cravant-les-Coteaux

www.bertrandsourdais.com

'Wine is my life. It's my way of expressing myself,' says Bertrand Sourdais. When I meet the fifth-generation winemaker at his Chinon property, he has just driven from his other winery – not in a neighbouring appellation, as one might expect, but in Ribera del Duero. It's quite a drive from northern Spain to the Loire Valley, but this is a journey he makes on a regular basis, enormous dog in tow, and he doesn't seem to consider this unusual. But then Bertrand Sourdais is not your average Chinon winemaker.

His education began traditionally enough at the local agricultural lycée in Montreuil-Bellay, where he was a contemporary of Antoine Foucault

(Domaine du Collier) and Sylvain Grosbois (Domaine Grosbois). He gained blue-chip experience in Bordeaux at Mouton-Rothschild and Léoville-Las Cases before heading to Chile for a spell at Santa Rita, and then later to Spain to work with Alvaro Palacios. Further studies and work in Bordeaux followed, and he then returned to Spain for a decade, where he developed Dominio de Atauta in Ribera del Duero.

Since coming back to take over from his father Jean-Bernard in 2005, Bertrand has made significant changes to the domaine, in both the vineyard and the winery, although the latter – a tiny, low-ceilinged, beamed cellar under the house – would probably not have looked out of place when his ancestors began farming this land in the late 1800s. Bertrand's vision for qualitative changes included conversion to organic and then biodynamic viticulture, and refining the selection of grapes from the family's 14 hectares to create a cohesive range of terroir-driven wines.

In 2007 he planted an area adjacent to the cellar with a high-density massal selection taken from old vines of the domaine, as well as a selection from Domaine des Ouches and Clos Rougeard. The fruit from these young vines currently goes into the entry-level Cabernet Franc cuvée. Bertrand also reimagined the historic blended house cuvée, Les Pensées de Pallus, in 2013, by limiting the fruit to a small selection from the best vines found on the hillside. This wine is fermented in concrete, aged for a year in 228-litre burgundy barrels and kept for three years in the cellar before release.

Speaking of the qualities of Cabernet Franc, Bertrand says, 'It's like Grenache. It can be *unbelievable* – or just "meh".' It goes without saying that he is aiming for the former, and for his wines to be *vins de garde*: 'I consider that Cabernet Franc is not at its best when young,' he comments. 'It shows all of its complexity with time.' Tasting wines crafted with such low yields as this (30 to 35 hectolitres per hectare), and with the long but gentle macerations he favours, it's easy to see what he means. A recently bottled Les Pensées de Pallus shows intense, black, brooding fruit and bitter chocolate flavours. Enjoyable now, it clearly has much more to give. Some of the oldest vines of the domaine go into Le Clos de Pallus. This is an early-ripening site in the lower part of the property. The wine can be closed in its youth but over time opens to reveal the classic fragrance and gentle sweetness of Loire Cabernet Franc from mature vines.

The top two cuvées are La Croix Boissée and La Rougerie. These are outstanding wines from tiny parcels of land that Bertrand understandably refers to as his *grands crus*, even if such a classification does not formally exist in the Chinon appellation. While both are on the best, higher slopes, sheltered by the forest above, they have quite different soil types and characters. La Croix Boissée vineyard is classic clay-limestone over tuffeau subsoil. Bertrand explains that to balance the strong character of the tuffeau soil with that of the Cabernet Franc grape, he allows the vines to grow to more than two metres in height here (much higher than elsewhere) to create, in his words, 'more modern wines with more fruit: aromas of red, fresh fruits, cherry notes, and a floral signature – and always with good acidity.' It's impossible to be indifferent to his wine from La Croix Boissée. This is a simply beautiful wine that will stop you in your tracks with its gorgeous, silky texture and fine fruit complexity. It personifies the elegance and depth of fine Chinon wine.

La Rougerie is a different beast altogether. From a vineyard planted on a narrow band of deep clay, it comprises partly old vines (a massal selection planted by Bertrand's grandfather in 1952) and partly a more

Bertrand Sourdais is passionate about his terroirs

recently planted high-density selection of Cabernet Franc from the best vines of the domaine. The old vines were chosen by his grandfather for their small, loose bunches and small berries. The combination of small berries, clay soils, low yields and skilled winemaking results in a wine of inky depth and power, with massive but velvety tannins and incredible savoury length in the mouth. This is a wine of pure class to last for decades – which is just what its creator has in mind.

The focus of Domaine de Pallus is the truly excellent range of reds, but there is also a tiny amount of white wine made, Le Coq de Pallus, which is one of the best examples of Chinon Blanc.

Domaine Grosbois

Panzoult

www.domaine-grosbois.com

This relatively recent winemaking venture is the work of brothers Nicolas and Sylvain Grosbois, a couple of globetrotting winemakers who finally decided to put away their passports and return to their Chinon roots. With origins dating back to the fifteenth century, this family estate was taken over by the brothers' parents in 1991. Nicolas was the first to return, in 2008, from a decade of travel in New Zealand and Chile. His brother followed in 2018 after 19 years away from home, when he gained experience in Chile, Oregon, Australia and Brazil. The brothers are proud of their commitment to polyculture, a little like the olden days of wine production. But with a wealth of experience of the world beyond the vineyards of Chinon, it's no surprise that their methods are reassuringly modern.

Only red wines are produced here, from 25 hectares of Cabernet Franc vines planted close to the winery, mostly within a single, walled parcel. The brothers have planted 60 truffle oaks behind the vines. Just beyond, a path winds into a small forest where 19 pigs forage to their hearts' content and 15 chickens live their best life in a spacious wooden hutch among the trees. Elsewhere are Black Angus cows, fields of cereals to feed them and a market garden where the local community can buy their produce – all organic, of course. It is also possible to stay on the estate in a renovated apartment in the heart of the vineyard, which comes complete with electric bikes to explore the surrounding area.

The brothers like to keep yields low and planting density high. Viticulture is biodynamic and winemaking an apparently straightforward affair. Hand-harvested grapes are destemmed, hand-sorted and

Sylvain Grosbois, happy with his wines

dropped into concrete tanks. And that's it (plus the decades of experience and expertise, of course). Sylvain says they prefer a wider shaped tank to allow more 'infusion' to take place. There is no crushing, he says, because 'I don't want to make a soup of grapes.' There is also a concrete egg, but barrels are noticeably absent. Here, it's all about the fruit, capturing the freshness, fragrance and purity of Cabernet Franc grown on the warm yellow '*millarges*', the local name for the sandy limestone tuffeau soils found here. Sulphur dioxide use is kept to a minimum and the wines are often within the limits of the classification Vin Méthode Nature (under 30 milligrams per litre).

La Cuisine de Ma Mère is a fun and inexpensive wine for early consumption. Macerated for six to ten days only, it is juicy and fresh, with expressive fruit. The signature cuvée of the domaine, Gabare, is named after the traditional flat-bottomed boats of the Loire. It is a blend of plots, from the yellow *millarges* on the slopes and gravelly soils towards the Vienne river. It shows typically bright fruit, but with a dark and silky, sooty depth. The wine matures for up to a year in concrete to preserve the purity of the fruit. This is an enormously enjoyable, balanced and accessible wine. The flagship cuvée, Clos du Noyer, comes from a small parcel of 1.24 hectares at the top of the estate, a site cooled by its proximity to the forest. The soil is thinner and more calcareous here, giving a more structured wine with fine tannins and cool, pure fruit. There is longer maceration, but again, no oak. A more upright wine, it is nevertheless deliciously drinkable even in its youth, though it will benefit from a few years in bottle. Clôture is an exceptional wine from a tiny plot of century-old vines in the middle of the hillside, planted in 1905 when the first rootstocks arrived from the United States following the phylloxera crisis of the nineteenth century. The first vintage of this wine for the domaine was 2018. This warm vintage produced a dense and fleshy wine that combines intensity and elegance, with the same fruit purity that characterizes all of these wines. A final cuvée is the Grosbois' sole concession to the oak barrel. Montet comes from a parcel of grapes whose roots plunge directly into the limestone rock, some 2 kilometres from the main vineyard. This gives a more austere expression of Cabernet Franc that is softened by ageing for 18 months in one- or two-year-old barrels. This is a departure from the usual house style, firmer and definitely in need of some time in bottle.

An air of joie de vivre pervades Domaine Grosbois and some of it seems to seep into the wines. Their lush texture, highly expressive fruit and flagrant deliciousness cannot help but raise a smile.

Domaine Olga Raffault

Savigny en Véron

www.olga-raffault.com

Although there had been vines in the family for generations, cultivated for wine alongside other crops, the Olga Raffault estate became better known when its eponymous founder and her husband Pierre focused on wine production in the 1930s. Today the family's 24 hectares of vines are in the safe hands of Olga's granddaughter, Sylvie, who manages the estate with her husband Eric de la Vigerie and their son Arnaud.

The domaine has been organic since 2015. Although sulphur usage is fashionably low, the Raffault wines show no trace of the funky (or faulty) flavours that often afflict those of domaines choosing to go down the more natural winemaking route. Attention to cellar hygiene helps keep these wines fresh and bright.

The grapes are picked parcel by parcel and held at cold temperatures in the winery until all of the harvest is in. Only then does fermentation begin. Les Barnabés, the entry-level cuvée, comes from lower-lying alluvial sites and shows soft, ripe, black fruit and primary floral aromas. Highly drinkable, this has a youthful appeal and lovely concentration. Les Peuilles comes from warm, south-facing flint and clay soils to the north of the city of Chinon. Sylvie says this gives the wine a 'soft, unctuous style'. Open-textured, with confit red and black fruits, it offers terrific value for money. Perhaps the best-known of the Olga Raffault cuvées is Les Picasses. Here, on one of the best sites in the appellation, calcareous soils give age-worthy structure to the wine, enhanced by ageing in large, old oak foudres. The hand-picked grapes are destemmed and the berries fermented whole at temperatures of up to 30°C, followed by a long maceration. This is a serious wine that ages beautifully. The La Singulière wine comes from a meticulous selection of grapes from the oldest vines of the domaine. It undergoes cold maceration for a month and ages for two years in oak, which gives the intense black fruit a hint of spice.

A warm welcome is assured at Domaine Olga Raffault

The domaine also produces two white Chinons (one aged in oak) and a deliciously ripe rosé that is one of the best in the area. This is a welcoming place to visit, and open to the public all year. Even better, it is possible to buy older vintages on-site, so the full potential of these wines can be enjoyed without the wait.

Domaine Philippe Alliet

Cravant-les-Coteaux

Philippe Alliet and his wife Claude took over the family domaine in the heart of the Chinon vineyard in the early 1980s with a clear mission to create wines of quality and in this they have succeeded, admirably. An early adopter of the idea of producing soil-specific cuvées, Philippe, now working with his son, Pierre, has built a formidable reputation. The Alliet wines are among the finest expressions of Cabernet Franc anywhere.

The family works 18 hectares of land organically. Every grape is harvested by hand, and one feels as if attention to detail is simply a way of life here, from the most basic cuvée to the top of the range. A small quantity of white is made but, as is the case throughout Chinon, it is the reds that provide the 'wow' factor.

The first wine in the range, Tradition, is, Pierre says, 'the most complicated to ripen.' Here, the grapes grow on the plain, on sandy gravels. 'The soils are less good quality, so we need many hours of work in de-budding and selection – we throw a lot away,' he says, reiterating later, as we taste it, 'You should see the selection we do for these wines: it's impressive.' Tradition is a hugely attractive wine, with smooth, spicy character and easy-drinking, gentle tannins. The cuvée Vieilles Vignes comes from vines on similar soils ranging in age from 50 to 85 years. This shows succulent concentration of bright fruit, and the effortless balance that comes from old vines. There is no oak used in either of these wines, so the quality of the fruit is plain to see. The Vieilles Vignes spends 18 months in concrete tanks, where it loses none of its violet freshness.

L'Huisserie comes from grapes grown on higher slopes of siliceous clay, planted in 2001. In the early years, Pierre says, the vines were very vigorous, leading to irregular vintages, and the wines were sometimes very high in acidity. But since 2015 the wine has been hitting its stride and will only improve as the vines come into balance with greater age. The flint in the soil means this wine will always have bright acidity lifting the dark, smoky fruit. A light touch of oak adds to its appeal.

The flagship cuvée is Coteau de Noiré, a single-hectare south-facing plot on clay-limestone planted in 1978. This is the only wine of the domaine where the oak is evident in youth, but this is a wine firmly intended for ageing. 'It's best to wait for 8 to 10 years before opening,' advises Pierre. Nonetheless, it is beautiful in its youth, with silky cassis and tobacco fruit, and fine tannins. Compact when young, it opens to reveal the purity, freshness and balance that come from the family's tireless dedication to quality.

Domaine Wilfrid Rousse

Savigny en Véron
www.chinonrousse.fr

'Don't get out, we're driving in,' says Wilfrid Rousse, as he hops out of his van to open the gates to his barrel cellar. He's not the only winemaker in the Loire to have the luxury of a drive-in cellar but it's a reminder of how vast these tuffeau cellars can be – and of the depth of the limestone that lies under the vineyards. In this silent, cathedral-like space, 15 metres below ground, the temperature is a constant 11°C, the air still and humid. 'It's a key part of the terroir,' says Wilfrid, adding, '*le vin ne prend pas une ride*,' meaning: the wine doesn't age in the slightest (literally, 'not even a wrinkle').

With no history of winemaking in his family, Wilfrid started from scratch. He studied viticulture and oenology in Amboise before buying his first plot of one-third of a hectare in 1987 at the age of 22. With no older generation to follow (or fight with), he has been free to make his own decisions and change his approach over the years. 'In the '90s, I did what I'd been taught,' he says. This involved fermenting in stainless steel ('it was fashionable in the '80s'), and many pump-overs. He still ferments in stainless steel, but these days moves the wine very little during fermentation, and only racks once during maturation. He has gradually reduced his sulphur additions to zero, preferring to rely on his ideal cellar conditions and, crucially, the fact that his young wines are saturated with protective carbon dioxide, having had little disturbance during fermentation. In short, he is doing less and less – although, as he says, 'The hardest job is to do nothing.'

The wines are bottled unfiltered, with a small dose of sulphur dioxide. (His son is making some sulphur-free wine, but one senses Wilfrid is not ready to risk 20 hectares of hard work on this just yet.) Les Galuches comes from vines close to the confluence of the Vienne and the Loire.

Bright and fresh, with crunchy raspberry and cranberry fruit, it sees no wood and is perfect for early consumption. Les Puys comes from a vineyard planted in 1990. It shows darker, more chocolatey fruit, with some black cherry and black pepper notes. The wine spends a year in oak, but is not marked by it. It is clean, fresh and delicious. Les Bois de Beaumont Vieilles Vignes is from three parcels of 50-year-old vines. Elegant, light but concentrated, this is simply lovely. Wilfrid comments, 'You see all of the fruit here, because we use so little sulphur [dioxide].' The top cuvée is Clos de la Roche, aged for up to two years in older oak. This needs time in bottle but shows silky texture and sweet, floral fruit.

This is a domaine doing what the Loire does best: producing effortlessly drinkable wines with bright, forward fruit, and at prices that won't break the bank.

La Niverdière

Beaumont en Véron
www.laniverdiere.com

Martine Budé is a woman who wears her heart on her sleeve – or at least, on the labels of her wine bottles. With names like Renaissance, Palimpseste and Resilience, she clearly has a very personal story to tell.

This is an historic domaine, but the more recent history of winemaking here dates back only to 2015 when Martine, the former wife of a Chinon winemaker (yes, there's the story) decided to try her hand at making wine herself. It turns out she's rather good at it.

Describing herself as a phoenix rising from the ashes to begin her second life, she purchased the Clos la Niverdière with 7.5 hectares of Cabernet Franc vines. Of this, 3 hectares are within the Clos, and 4.5 hectares are in the commune of Cravant les Coteaux. She has converted the estate to organic viticulture and says she also works a lot with biodynamic practices in the vineyard. All of the fruit is hand-harvested and fermented with indigenous yeasts in concrete vats.

Her Renaissance cuvée comes from vines planted in 1970 on gravel and sand alluvial soils close to the Vienne river. It is supremely silky and elegant, with very supple tannins even when young. Palimpseste comes from the vines planted at Clos La Niverdière in Beaumont en Véron, which grow on sandy clay over limestone. This is another gentle, pretty wine, with slightly more structure and mineral acidity. Both have lightness and purity of character. Martine tells me she extracts tannins by pumping over, but to only a limited extent. 'I don't like strongly extracted tannins,' she says, 'I want it to be easy to drink and easy to understand.' Resilience is the most tannic and ageworthy of the three cuvées. Grown on the clay-limestone soils of Cravant les Coteaux, this is a more serious wine, made in the same way, that shows a higher volume of fine-grained tannins and lifted aromas of violets and dark chocolate. It's delicious.

From tasting early vintages of the domaine, there are some lovely wines here, but one feels the best is still to come. Here's to new beginnings.

Other Chinon producers to try

Domaine de Noiré, Domaine de la Commanderie, Château de Coulaine, Domaine Jourdan, Domaine de la Marinière, Vignoble du RAB, Jean-Maurice Raffault.

Bourgueil and Saint-Nicolas-de-Bourgueil

These two appellations, on the north bank of the Loire, are separate appellations, but in all other respects are closely linked. On a map, the vineyard area of Saint-Nicolas-de-Bourgeuil – the Loire's only single-commune appellation – looks like a slice of pie extracted from the larger appellation of Bourgueil, which covers seven communes. The area has a long wine-growing history, the vineyards having been developed by the monks of the abbey of Bourgueil more than a thousand years ago.

The appellations are situated just to the north of Chinon, across the Loire. The climate is a little cooler here but, as with their neighbour to the south, the vineyards of Bourgueil and Saint-Nicolas are protected by trees to the north, which block cold winds and moderate the climate. The vines in the lower-lying Saint-Nicolas sites enjoy a slightly warmer climate and tend to have lower acidity. The tuffeau slopes, which make up around half of the Bourgueil appellation, give more structured and complex wines.

Both were granted appellation status at the same time in 1937 and, as neighbouring communes producing similar styles of wine, the growers' Syndicat originally included both. But over the years, local rivalries evolved based on the idea that the single-commune Saint-Nicolas was assuming superior status to the larger Bourgueil, as if Bourgueil was a generic appellation and Saint-Nicolas a superior part of it. This is a curious idea, firstly since the only reason there are two appellations in the first place is seemingly because the villagers of Saint-Nicolas were quicker off the mark to apply for their own appellation status back in 1937. And secondly, the terroir of Saint-Nicolas is almost identical, or arguably slightly inferior to that of its larger neighbour, since it comprises mostly the flatter, deep, sandy soils and less of the tuffeau slopes where the more 'serious' wines are grown. While Saint-Nicolas is appreciated for its youthful, fruity drinkability, does this mean the wines are better? It is a more densely planted area and, because it has slightly higher permitted yields, tends to produce more wine. But feuding aside, the most important quality indicator for the consumer is, as ever, the name of the producer on the bottle – even if, confusingly, many of them share the same name here. As the producers of both appellations are scattered across the entire area, and many producers make wines in both appellations, they are grouped together below.

Bourgueil – key facts

- **AOC granted**: 1937
- **Hectares in production**: 1,183
- **Annual production volume**: 49,000 hectolitres/6.5 million bottles
- **Grapes**: Cabernet Franc, usually 100 per cent (Cabernet Sauvignon is permitted up to 10 per cent)
- **Styles**: 96 per cent red, 4 per cent rosé
- **Soils**: Two main soil types. The vineyards closer to the river comprise alluvial sand and gravels; those on the slopes are tuffeau
- **Permitted yield**: 55 hectolitres per hectare
- **Communes**: seven – Restigné, Bourgueil, Benais, Ingrandes-de-Touraine, Saint-Patrice, Chouzé-sur-Loire, La Chapelle-sur-Loire
- **Winemakers**: 120

Saint-Nicolas-de-Bourgueil – key facts

- **AOC granted**: 1937
- **Hectares in production**: 1,059
- **Annual production volume**: 56,500 hectolitres/7.5 million bottles
- **Styles**: 99 per cent red, 1 per cent rosé
- **Grapes**: Cabernet Franc, usually 100 per cent (Cabernet Sauvignon permitted up to 10 per cent)

- **Soils**: Two soil types: two-thirds of the land is alluvial terraces of deep sand and gravel soils. One-third is on higher slopes of tuffeau covered with sand.
- **Permitted yield**: 58 hectolitres per hectare
- **Communes**: Saint-Nicolas-de-Bourgueil
- **Producers**: 96, all independent

Notable Bourgueil and Saint-Nicolas-de-Bourgueil producers

Domaine Amirault – Clos des Quarterons

Saint-Nicolas-de-Bourgueil
www.domaineamirault.com

Xavier Amirault says his family has been in the region for 200 years, which is no surprise when you see how many Amiraults there are in this little village today. He is the sixth generation to make wine, based at the family's historic home, the Clos des Quarterons, in the heart of Saint-Nicolas-de-Bourgueil.

This is a welcoming domaine that encourages visitors. It is also very evidently a nature-focused domaine, where you are greeted by a courtyard overflowing with aromatic herbs and flowers, some for cooking (they do a mean barbecue here), some decorative and others for the biodynamic plant preparations used in the vineyard.

The estate comprises 38 hectares, with vines spread over 58 plots, mostly in Saint-Nicolas-de-Bourgueil, but a few in Bourgueil. A wide range of wines is made, including a crémant. Xavier is passionate about biodynamics, seeing it as a way of life rather than just a way of making wine, and it underpins all of the activities of the domaine. Some of the wines are also certified as Vin Méthode Nature, meaning that they are made with very little or no sulphur dioxide. For me, these latter wines are the least interesting the domaine has to offer, as they can be tainted by the usual problems that afflict natural wines: oxidation, *Brettanomyces*, or other off-flavours. Happily, there are many other wines to choose from here.

The crémant wine, Armantine, is a blend of 70 per cent Chenin and 30 per cent Chardonnay. It spends 24 months or more on lees and is creamy, saline and elegant. The 'house cuvée' is Les Quarterons. Representing 30 per cent of the domaine's production, it is from a blend

The entrance to the cavernous cellars of Domaine Amirault

of plots of sandy soils. Juicy and fresh, this easy-going, early-drinking, red-fruited Cabernet Franc style is, says Xavier, exactly what they're aiming for. It is a perfect example of well-made, modern Saint-Nicolas-de-Bourgueil. The Clos des Quarterons Vieilles Vignes is produced from grapes coming from vines that are at least 40 years old. This has 15 months of oak ageing. The wine has more concentration and could be kept for up to 5 years. Another notable old vine cuvée is Le Fondis, which comes from 65-year-old vines that grow on gravelly terrain. Firm in its youth, it takes a year or two to reveal its charms. Le Vau Renou is the top Saint-Nicolas cuvée of the domaine. It is a perfumed, supple and spicy wine that ages for 30 months in part-buried amphorae in the vast cellar that lies under the vineyard itself. It's worth visiting the domaine just for a sight of this beautiful cellar.

Domaine de la Chevalerie

Restigné

www.domainedelachevalerie.fr

Laurie and Emmanuel Caslot are the thirteenth generation in their family to be working the soils of Bourgueil, as their ancestors have done since 1640. The siblings manage together the 33-hectare estate, situated almost all in a block around the winery on the slopes of Restigné and Benais. All of the wines are in the Bourgueil appellation.

Biodiversity is an important part of the biodynamic philosophy here. They plant cover crops to help soil structure and have permanent grass growing in the vineyards. In some years this can hinder ripening, Laurie says, as the grass competes with the vines, so they crush the grass periodically rather than mowing it. They have also planted trees and created hedgerows to preserve and enhance life beyond their vines.

It was as recently as the 1980s that the family changed from polyculture to wine alone. The wines are expressions of the multiple terroirs. The more immediate, short-term wines come from parcels lower down the appellation towards the Loire. Branded with fun names such as Franco de Porc, and Noms d'Oiseaux, these '*vins de copains*' – wines to enjoy casually among friends – are unpretentiously joyful, all smooth succulence and deliciousness. 'To enjoy under a beautiful starry sky or a relaxed barbecue,' suggests the label, and for this, they are just about perfect.

For the 'terroir' wines, the labels look a little more sober, but the wines are no less fun. There's the fragrant and fruity Peu Muleau (meaning 'little hill') from sandy soils towards the top of the appellation, and the more assertive but elegant Galichets, from a site with a little more clay. Chevalerie is more of a keeper. It has more complexity and finer aromas but retains the impression of lightness that characterizes all of the domaine's wines. Both of the latter two wines are from old vines of more than 60 years of age. The most long-lived cuvées of the domaine, Bretêche, Busardières and Grand-Mont, come from the best chalky clay soils of the appellation. These are wines that can age for 10 to 15 years, but are approachable in their youth.

There is a consistent style to the wines of Domaine de la Chevalerie, which are made with care and restraint. Yields are low, averaging 30 to 35 hectolitres per hectare (the permitted yield for the appellation is 55 hectolitres per hectare). The grapes are all hand-harvested, fermented with indigenous yeasts and gently handled. 'We are not looking for great tannic extraction,' says Laurie Caslot, 'we like finesse.' The wines are bottled unfiltered and then left to rest for a couple of years in the winery to continue their development before release, so the wines are sold when they are good to drink.

Domaine du Bel-Air

Benais

www.domainedubelair-bourgueil.fr

Pierre Gauthier is, in his own words, '*un pur cru de Bourgueil*'. Born 50 metres from the winery, his destiny as a winemaker seems never to have been in doubt. The history of this family domaine is relatively short, though (in French terms, at least). Domaine du Bel-Air dates back to the late 1800s, when Pierre's great-grandfather, a potter, decided to turn his hand to vegetable growing and bought some land. He later took advantage of the inexpensive land available after the phylloxera crisis and planted a few acres of vines that he grafted himself onto the newly available American rootstocks. The domaine grew from there and today Pierre and his son Rodolphe farm 60 parcels across 20 hectares around Bourgueil.

With no formal winemaking education, Pierre learned his trade the old-fashioned way: from his father. 'I'm no revolutionary,' he says, 'if people have been making wine this way for a thousand years, I'm not going to do things differently.' Nevertheless, things have changed since his parents' day. 'Forty years ago, we didn't do much; now we are much more active in the vines,' he says, adding, 'The vignerons here are like gardeners. They really love their vines.' And despite his modest protestations that he just followed in his father's footsteps, it is clear that he has spent a lifetime paying close attention to his vines and his wines, developing a knowledge that he is now passing on to his son, who has officially been managing the domaine since 2016.

The grapes are almost all hand-harvested before being destemmed and macerated at cold temperatures for a few days before fermentation in stainless steel. They use natural yeasts, and temperatures are kept at a relatively cool 27–28°C ('better for the yeasts, better for the aromas,' says Pierre).

Five cuvées are made, all 100 per cent Cabernet Franc and all are AOC Bourgueil. Jour de Soif is the appropriately named entry-level wine, a *vin de soif* that doesn't disappoint, with pristine, bright, raspberry fruit. Les Vingt Lieux-Dits also does what it says on the tin: a blend of twenty parcels of vines that grow on clay soils and make a wine that, according to Pierre, 'represents Bourgueil.' It has lifted aromatics, smoky, juicy black cherry and blueberry fruit, a veil of crisp, crunchy tannins and a peppery edge. Tasting the 2020 vintage, Pierre remarks, 'In the past, we could only make a wine like this every 20 years. Climate change has been good for Cabernet Franc.' The Les Marsaules wine comes from the domaine's oldest vines, planted on silica-clay soils. It ages for 30 months in one-year-old 500-litre barrels. Mouthwateringly fresh in its youth, it shows blackcurrant fruit and elegant, fine tannins. Even in a warmer vintage, there is no sensation of overripeness because, as Pierre points out, 'we never lack acidity here'.

Grand-Mont is a serious wine that comes from clay-limestone soils at the top of the appellation. It spends 30 months in oak, and shows toasty, complex, savoury fruit intensity, the concentration balanced with a chalky minerality. Clearly a *vin de garde*, it has a characteristic freshness that makes it approachable even in its youth, a reflection I

Pierre Gauthier realized his dream in buying Clos Nouveau

share with Pierre. 'Nobody's going to be angry if you open it now,' he says, with a chuckle.

The *grand vin* of the domaine is Clos Nouveau, from a hectare of vines within a seventeenth-century *clos* that had been left to abandonment before the Gauthiers bought it. This is an early-ripening site and the wine is in a different league: simply gorgeous, with silky texture, rich, black fruit and an endless finish. It has masses of substance but no sensation of weight, a wine that will last for decades. Pierre recalls that he would walk past the gated *clos* every day on his way to school. 'At that time, people would say, "this is where the good wine is",' he recalls, 'yet when it came up for sale in 2005, nobody wanted it; they were all coming up with reasons not to buy it, so we just quietly put in an offer and bought it. Nobody could believe it, because those things are for rich people, not for the likes of us.'

Domaine Yannick Amirault

Bourgueil

www.yannickamirault.fr

Concern for the environment has always been paramount for Yannick Amirault. Rejecting the vogue for heavy use of agrochemicals in the 1970s, he chose to set up on his own instead of taking over his father's vines, and in 1977 began farming 3.9 hectares of the Les Malgagnes vineyard belonging to his grandfather in Saint-Nicolas-de-Bourgueil. Over the years he, like many others, has turned back the clock, returning to more traditional methods in the vineyard and winery, and achieving organic certification in 2009. Yannick was a pioneer in doing what is now commonplace among good producers: planting grass between the rows, and cover crops over the winter. The Amiraults refer to their viticulture as gardening, and follow the lunar calendar in their activities.

Yannick's son Benoît joined the family business in 2003 and today the domaine counts 19 hectares across 30 parcels, of which two-thirds are in Bourgueil and one-third in Saint-Nicolas. Benoît says the difference between the two appellations is that the predominantly gravelly soils of Saint-Nicolas are flatter, warmer and more free-draining, compared to the cooler slopes of Bourgueil, where there is more clay. He offers his own take on the popularity of Saint-Nicolas: 'It's easier to sell anything with "Saint" in the name.'

The pristine cellar is equipped with large conical wooden vats that receive the hand-harvested grapes for fermentation. Ageing is done

Barrels and amphorae at Domaine Yannick Amirault

in barrels, or 600-litre sandstone amphorae ('more neutral than terracotta,' Benoît says). The macerations are very gentle here, the aim being to achieve infusion rather than extraction of tannins, especially for the lighter wines in the range. This is evident in the wines, which are fruit-focused, clear and bright. Seven reds are made, three from Saint-Nicolas and four from Bourgueil. The stand-out cuvées for me are the ripe, perfumed Bourgueil Le Grand Clos, the intense and elegant Saint-Nicolas-de-Bourgueil Les Malgagnes and the sophisticated Bourgueil Le Pavillon du Grand Clos. But there is plenty to enjoy throughout the range. The entry-level La Source from Saint-Nicolas shows a generous, open nature, and is perfect for early consumption. Even at this level, the vines are more than 40 years old. The domaine also produces a gorgeously creamy, lees-aged barrel-fermented Rosé d'Equinoxe and a barrel-fermented white wine from Chenin Blanc, Bâtard-Princesse, that performs well above its IGP Val de Loire status (there is no white appellation here).

Yannick Amirault says, 'What counts in Bourgueil is drinkability. You don't want to notice the bottle emptying.'

Vins Cognard

Chevrette, Saint-Nicolas-de-Bourgueil

www.vins-stnicolas-bourgueil-cognard.fr

This family domaine is run almost singlehandedly by Estelle Cognard. It was started by her parents, Lydie and Max, in 1974, when Lydie inherited a sole hectare of vines from her parents, who grew a few vines alongside their pear and apple trees.

Over the years the business has grown to 15 hectares – 11 in Saint-Nicolas-de-Bourgueil and 4 in Bourgueil. Estelle joined the business in 1997 and her younger brother, Rodolphe, a few years later, taking over the winemaking side of the estate. The domaine improved from year to year, as they moved towards organic viticulture, and built a temperature-controlled cellar. And then came a tragic loss for the family: Rodolphe died in 2018. His sister has carried on alone, now using a winemaking consultant to assist with the vintage each year.

There are five red cuvées, plus a little rosé. All are fermented in temperature-controlled stainless steel. The most recently created cuvée is Cuvée DOR, created in memory of Rodolphe in 2018. This is a youthful, early-drinking wine that is fermented in stainless steel and macerated for 10 days. It is all juicy summer fruits, and very easy to enjoy. Cuvée Estelle comes from similar soils, mostly sandy, but on slopes. This is macerated for a little longer, around two weeks. It has more substance and a pure raspberry fruitiness. Les Tuffes comes from a mixture of terroirs and is one of two reds produced in the Bourgueil appellation. The vines in the village of Benais grow on clay-limestone, while the ones in Bourgueil are on sandier soils. It is aged in old oak for 10 months and has masses of black fruit appeal and more structure than the previous two wines. The flagship cuvée is Les Malgagnes, named after the *lieu-dit* of the same name, where they own 3 hectares. This is a fine terroir, turning out lovely balanced wines year after year. Created in 1976 by Lydie and Max, the wine spends a year in wood, and is dominated by its stylish silky fruit. The domaine's top wine is Les Caudalies, which grows on clay and limestone and spends up to 18 months in old oak. This is floral and elegant, with a concentrated core of black fruit.

The Cognard wines are easy to love and offer excellent value. The domaine is open for visitors all year (although it is best to call ahead) and wines are available to buy at the domaine, including older vintages.

Other Bourgueil producers to try

Aurélien Révillot, Frédéric Mabileau, Domaine de la Lande, Lamé Delisle Boulard, Château de Minière, Domaine de la Cotellerie, Domaine de la Noiraie.

When in the area

Places to stay

In Chinon, a good base for exploring is the Hotel Diderot (www.hoteldiderot.com), a small and friendly family establishment run by Jean-Pierre Dagneaux and his American wife Jamie. Clad in jasmine and roses, the traditional stone building close to the centre of town dates back to the fifteenth century. Rooms are characterful and cosy, and there is a bar to sample local wines. Breakfast around the old stone fireplace features Jamie's exceptional home-made jams.

Tarte vigneronne tastes as good as it looks

Places to eat

Away from the more touristy restaurants of the central square in Chinon, Au Local (www.aulocalrestaurant.fr) is worth a visit. It features modern French cuisine from an enthusiastic young couple, and you can sip a glass of Chinon while admiring the castle. It's situated above the town, directly opposite Couly Dutheil's Clos de L'Echo vineyard. In the summer of 2024, Domaine Grosbois launched a seasonal outdoor restaurant at the winery, La Table du Pressoir (www.domaine-grosbois.com/restaurant-la-table-du-pressoir). From its elevated position, diners have an expansive view across the vines and river. The menu features meat, vegetables and wine all grown on the estate. For a special meal, the Michelin-starred Les Années 30 (www.lesannees-30.com) has the added benefit of being one of the few restaurants in the region to be open on a Monday. And don't miss the opportunity to try a local Chinon speciality, the delicious *tarte vigneronne*. This is a thin and crisp puff pastry base topped with apples cooked in Chinon wine, which makes it as good to look at as it tastes. Find it at Pâtisserie Ayrole in Chinon, where the recipe was conceived. In Saint-Nicolas-de-Bourgueil, try Chez Odette, owned by the Mabileau family (www.fredericmabileau.com/fr/restaurant), in the main square.

THE WINES OF THE LOIR: JASNIÈRES, COTEAUX DU LOIR AND COTEAUX DU VENDÔMOIS

The three most northerly appellations of the Loire Valley follow the course of the Loir, a river that meanders from east to west in an approximate parallel course to the Loire. Despite the proximity of the name, it is not a direct tributary of the Loire; north of Angers, it empties into the Sarthe, which shortly afterwards joins the Mayenne to form the Maine, which then flows into the Loire just south of Angers.

There's quite a contrast between the scale and power of the Loire and this little river. If the Loir is going anywhere, it's in no hurry to get there. In winter, it's a very quiet place to be and even in summer, it's a world away from the tourist hordes. The vineyards of the Loir are generally quite well hidden. It is a quiet and rural area, with the odd grand private château sitting resplendent along the banks of the sleepy river. Towards the historic city of Vendôme, the only place of any significant size in this area and a mere 45-minute TGV train ride from Paris, there are more obvious signs of human activity, such as fields of cereals, cement works and intermittent vineyards.

Jasnières – key facts

- **AOC granted**: 1937
- **Hectares in production**: 75
- **Annual production volume**: 2,500 hectolitres/330,000 bottles
- **Styles**: 100 per cent white
- **Grapes**: Chenin Blanc
- **Soils**: flinty clay over tuffeau limestone

Coteaux du Loir – key facts

- **AOC granted**: 1948
- **Hectares in production**: 76
- **Annual production volume**: 2,400 hectolitres/320,000 bottles
- **Styles**: 47 per cent red, 38 per cent white, 15 per cent rosé
- **Grapes**: Chenin Blanc, Pineau d'Aunis, Cabernet Franc, Côt, Gamay, Grolleau
- **Soils**: tuffeau and flinty clay

Coteaux du Vendômois – key facts

- **AOC granted**: 2001
- **Hectares in production**: 94
- **Annual production volume**: 3,700 hectolitres/500,000 bottles
- **Styles**: 40 per cent red, 32 per cent rosé, 28 per cent white
- **Grapes**: Chenin Blanc, Chardonnay (up to 20 per cent), Pineau d'Aunis (50 per cent or more), plus Cabernet Franc, Pinot Noir, Gamay
- **Soils**: flinty clays

This is a small and marginal area for grape-growing. In the western half are the appellations of the Coteaux du Loir and Jasnières, and to the east are the vineyards of the Coteaux du Vendômois. Jasnières is a bijou appellation for white wines only, made from Chenin Blanc. It occupies the upper north-east corner of the Coteaux du Loir vineyard and covers just two villages, Lhomme and Reuillé. The land in the Jasnières appellation is unusual for its humidity, which comes from the forest of Bercé to the north and the network of rivers that feed into the Loir. The presence of the forest blocks the humidity from the rivers and contributes further to this humid environment. This creates a unique microclimate that gives the wines added roundness compared to similar styles in neighbouring Coteaux du Loir. Across these two appellations combined there are 30 producers.

The Coteaux du Vendômois numbers even fewer growers, just a handful of individuals who often practise a polyculture of cereals, cattle and chickens alongside their vines. Almost 90 per cent of the vineyards around the city of Vendôme were pulled up after the Second World War. There has been some renewal of interest in grape-growing since the Coteaux du Vendômois was recognized as an appellation in 2001, but the great majority of production goes to the local cooperative. Ariane Lesné, president of the growers' association, says that before phylloxera, there were more than 2,000 hectares planted here. It's a familiar tale. 'Before we got the appellation yields were high and the grapes were usually made into rosé,' she says, 'but when yields were reduced (to 55 hectolitres per hectare) we realized we could make good reds.'

The red wine speciality of the Loir is Pineau d'Aunis, an historic grape variety found almost nowhere else. When yields are managed and vines are old, this variety is capable of producing pale, perfumed, but often powerfully tannic wines that may be light in body and colour, but are certainly not shy. They have a distinct personality with a peppery aftertaste and, when its tendency to overproduce is curbed, the wines have unique appeal. With growing curiosity from sommeliers and journalists interested in Pineau d'Aunis as a red wine with potential for quality, the growers in the Coteaux du Vendômois are increasingly interested in making wines with 100 per cent Pineau d'Aunis under the appellation. Traditionally, it has been more often vinified as part of a blend, which can also include Cabernet Franc, Pinot Noir or Gamay. But the challenge of Pineau d'Aunis here, says Ariane, is the tannins: 'The deep clay soils give high tannins, and you have to be very careful not to extract excess tannins from this grape because, once in the wine, they never disappear.' The growers of the Loir also produce palatable rosé wines – more often known as *vins gris*, (less appealingly in English, 'grey wines') – from Pineau d'Aunis.

Notable Loir producers

Domaine de Bellivière

Lhomme

www.belliviere.com

Wine lover Eric Nicolas was working in electronics for oil giant Total when he decided enough was enough. 'He was obsessed by wine,' his wife Christine recalls, 'so he went back to school to study oenology in Montpellier, before working in a cooperative in the south to get some experience.' But the couple had their sights set on more ambitious horizons. In 1995 they managed to buy a farm in the Loir, chosen as a halfway house between her roots in the south, and his in the north. With no vines of their own they rented a couple of hectares before gradually acquiring plots themselves. Today the domaine stands at 18 hectares, spread across 65 parcels of vines in five communes in the Jasnières and Coteaux du Loir appellations. From the 2021 vintage their son, Clément, has taken over the winemaking and runs the business with his wife, Laure-Anne, and Christine.

Even if the vineyards are well dispersed, the winery, cellars and house are all conveniently in one place, with naturally air-conditioned cellars dug into the rock behind the house. In 2015 Eric and Christine built a new, gravity-fed *chai*, with the grapes transported by elevator to the first or second floor. The top floor is for reds, which are all from Pineau d'Aunis. The hand-harvested grapes are destemmed before going into concrete tanks to ferment. Here they will undergo *pigeage à pied*, where the floating cap is pushed back down periodically by that gentlest of tools, the human foot. Later, the young wines will age in 350- or 500-litre oak barrels. Fruit from the oldest vines ferments in a wooden vat. On the first floor the Chenin arrives for gentle pressing in a pneumatic press and cold settling, before it is transferred to the cellars to ferment in oak with indigenous yeasts. Here the wines stay for up to two years.

The domaine produces multiple cuvées, which have been created to reflect the age of the vines rather than individual sites because, says Christine, 'there are just so many parcels.' The Chenins from Jasnières and Coteaux du Loir, she says, grow on similar, south-facing slopes and in very similar soils, with a bedrock of tuffeau underneath and upper soils of flinty clay, a combination that gives the wines added freshness. The main difference between the two appellations is the microclimate, she says, which is more favourable in Jasnières to the production of sweeter wines, especially those with botrytis.

Tasting the domaine's white wines, the Coteaux du Loir Prémices Sec Tendre from vines under 25 years of age is frank, fresh and bright. A Jasnières Les Rosiers Sec Tendre from slightly older vines (up to 35 years old) has a little more sweetness and complexity, with herbal notes and chalky tannins. L'Effraie Sec is a Coteaux du Loir from vines of a similar age. It is more upright and, as Christine comments, 'more monolithic'.

An older vine cuvée of Coteaux du Loir, Vieilles Vignes Eparses, comes from vines aged between 80 and 100 years old. It can be sweet or dry, depending on the vintage; either way, it is a terrific wine that shows restraint and delicacy. 'That's the interest of old vines,' says Christine, 'neither the sugar nor the alcohol shows.' And she's right, even though, at a maximum of 13.5% abv, these are by no means ever heavy wines. Their top, oldest-vine cuvée for Jasnières is Caligramme, which comes from 80-year-old vines. This has intense, savoury and slightly herbal fruit, with the honey of Chenin in the background. Christine advises, 'This is to open after 4 to 5 years, but you can keep it for 20.'

The main red of the domaine is Rouge-Gorge, a Coteaux du Loir that is pale, fragrant, light and characterful, with soft red fruit that is almost Gamay-like. 'Pineau d'Aunis is very well adapted for here,' says Christine, 'because it ripens even in difficult years compared to Cabernet Franc, which doesn't. But,' she adds, 'you need to watch the tannins.' This is the reason all the grapes are destemmed ('the stems are green') and extraction is as gentle as possible. The other consideration, of course, is that here we are a long way north, even for the Loire. 'There's always a vintage effect here,' says Christine.

One unusual and delicious wine that has only been made three times – the first by mistake, in 2001, because they forgot to harvest the grapes – and then again in 2005 and 2022, is Aurore d'Automne Coteaux du Loir Liquoreux, a sweet red wine made from Pineau d'Aunis. It is quite unlike any other sweet wine I've tasted. During harvest, the pickers leave two bunches of grapes on each vine to ripen further. The overripe, shrivelled grapes are then pressed and vinified as a white wine, in large barrels. Clearly, removing the skins is a good way to tame the tannins of Pineau d'Aunis as they are gentle, present but subtle here, adding complexity rather than structure. The wine has masses of red fruit, with a luscious honey and marmalade character and a surprisingly fresh acidity. The 2022 has 134 grams of residual sugar. It's absolutely delicious, but currently only available to buy at the domaine – every 17 years or so.

Domaine Colin

Thoré-la-Rochette

www.domaine-colin.fr

Certainly the most important producer of quality wines in this small appellation, the Colin family domaine is also the oldest; today,

Pierre-François is the ninth generation to be making wine on the gentle slopes of the Coteaux du Vendômois. The domaine is in the tiny village of Thoré-la-Rochette, at the heart of the appellation. There are 28 hectares under vine here, including 12 hectares of Pineau d'Aunis, 10 hectares of Chenin Blanc, 2.5 hectares of Pinot Noir, a hectare each of Gamay and Cabernet Franc, and 1.5 hectares of Chardonnay.

The minimum planting density for vines in the appellation is 4,500 plants per hectare, but here it is a far higher 7,500, in an attempt to increase competition in the vines and concentration in the berries. Maybe this is the secret, or perhaps it's the generations of winemaking expertise, but either way, these wines are at the top of the tree in this little-known outpost of the Loire Valley.

The range starts with an impressively bright and mineral Chenin, Pierre à Feu, that grows on flint soils. One of the advantages of having so many generations that have come before is the quantity of old vines on which to draw for mature and balanced fruit. One example is the savoury Blanc Vieilles Vignes, a Chenin from vines that are 60 to 80 years old, aged in barrels. The entry-level red, Pierre-François Rouge, is a blend of Pineau d'Aunis, Pinot Noir and Gamay. It is light, fresh, highly drinkable and good value. The Rouge Vieilles Vignes (they are 50 to 70 years old) is juicy but more structured, a blend of Pineau d'Aunis and Cabernet Franc. The two most interesting wines of the domaine come from old vine cuvées that are 100 per cent Pineau d'Aunis. The first is Rouge les Vignes d'Emilien Colin, named after the heroic ancestor who planted these vines between 1900 and 1920, after phylloxera. All I can think as I taste this is how incredibly moving it must be to be able to drink a wine made from fruit you've picked from a plant that your great-great-grandfather planted more than a hundred years ago as a new beginning when all seemed to be lost. Does it even matter what it tastes like? It is good, though. Incredibly, there are vines that predate even these, from a tiny parcel that is used to make a wine called Intuition. These precious grapes undergo fermentation in barrels of 600 litres. The wine is then left for 3 to 4 months before pressing and returning to barrels for a year. Pierre-François says the yields for this cuvée are only around 25 to 30 hectolitres per hectare. This is a *vin de garde*, although it shows a sweet ripeness to the tannins along with its minty freshness. It is a wholly harmonious expression of Pineau d'Aunis that can be enjoyed now, or perhaps by the tenth generation of Colins.

Other Loir producers to try

Domaine de Cézin, Cave du Vendômois.

CHEVERNY AND COUR-CHEVERNY

They may be a little off the beaten track in wine terms, but Cheverny and Cour-Cheverny receive more than their fair share of visitors. For this is prime Loire château country, where the formerly unacceptable excesses of past kings are now celebrated as national monuments, drawing crowds from far and wide. The grand châteaux of Blois, Chambord and Chaumont are among the more famous names gracing the vicinity, but you can't drive very far in these parts without catching sight of similarly impressive edifices looming up from the landscape, often seemingly out of nowhere. The Château of Cheverny is one such example. From a rural roundabout it swings majestically into view, the harmonious proportions of its seventeenth-century façade visible behind a gated, tree-lined alley. It's quite a distraction if you are behind the wheel.

The vineyards of Cheverny and Cour-Cheverny are intermingled and lie scattered across a broad area just south of the Loire, opposite the town of Blois. Here, at the eastern edge of the Touraine region, temperatures are on average up to 1°C lower than further downstream. This is due to the cooling influence of large forests to the east and south-east (the Sologne) and the river valleys of the Bièvre, Beuvron and Cosson, three small tributaries of the Loire. Vines grow on a diverse range of soils that are mostly sand-based but include stony limestone and flinty clay.

These are recent AOCs, promoted from VDQS in 1993. Cheverny is much the larger of the two. It produces mostly whites, which must comprise at least two varieties. These are typically easy-going but quite full-bodied Sauvignon-dominant wines, usually rounded out with fleshier Chardonnay. Menu Pineau (Orbois) may also be used. Reds and rosés are made from Pinot Noir blended with Gamay, with occasional additions of Cabernet Franc and Côt. These are more variable than the whites, but the best have juicy fruit, delicacy and finesse.

Cour-Cheverny accounts for a tiny proportion of the wines of the area, but it is renowned for its unique grape variety, Romorantin. Grown nowhere else in France (and possibly the world), locals claim it was introduced in 1519 by the famously prolific château builder, King François I, in Romorantin, which is where his mother used to live. Legend has it that he originally intended to build his palatial hunting lodge here instead of at Chambord, but was put off by a local plague infesting the area at the time. Cour-Cheverny wines are always white and all are 100 per cent Romorantin. Not for the faint of heart, these are a breed apart from their palate-friendly neighbours in Cheverny. Fiercely acidic Romorantin is not an easy grape. Naturally productive and late to ripen, it never loses its searing acidity yet oxidizes easily. A range of different styles can be found, not all good, from more oxidative examples to excessively oaky ones. But the top producers craft excellent and unusual, savoury, juicy and honeyed wines from this ancient variety, destined to last for decades.

Notable Cheverny and Cour-Cheverny producers

Domaine des Huards, Domaine de Montcy, Domaine Cazin.

Cheverny – key facts

- **AOC granted**: 1993
- **Hectares in production**: 700
- **Annual production volume**: 29,200 hectolitres/3.9 million bottles
- **Styles**: 65 per cent white, 28 per cent red, 7 per cent rosé
- **Grapes**: whites – Sauvignons Blanc and Gris (60 to 85 per cent), plus Chardonnay, Chenin Blanc, Menu Pineau (Orbois); reds and rosés – Pinot Noir (60 to 95 per cent), plus Gamay and Côt
- **Soils**: sand and clay, some flint, limestone and gravel terraces

Cour-Cheverny – key facts

- **AOC granted**: 1993
- **Hectares in production**: 65
- **Annual production volume**: 3,000 hectolitres/420,000 bottles
- **Styles**: 100 per cent dry white
- **Grapes**: Romorantin
- **Soils**: sand and clay, some flint, limestone and gravel terraces

When in Cheverny

Other than the château, there is not a lot else to see in tiny Cheverny, but wine lovers would do well to make a beeline for the well-appointed Maison des Vins which (as the more revolutionary-minded visitor will note) is literally a stone's throw from the château itself (www.maisondesvinsdecheverny.fr). Created by the winemakers of both appellations, it offers an original and immersive 360-degree artistic presentation of the region and the annual work of its winemakers. There is also a terrific tasting room here, where visitors can access up to 100 different wines from an ingenious, locally designed system that uses microchipped glass technology to automatically deliver a tasting measure from a spout connected to each bottle, which is then topped up with inert gas. The canny winemakers of these twin appellations have also installed a second Maison des Vins at the colossal and hugely popular Chambord Château, close by.

AOC TOURAINE AND THE LESSER KNOWN APPELLATIONS

At over 100 kilometres long, the Touraine AOC is one of the Loire Valley's largest appellations. Bisected by the Loire, flowing from north-east to south-west, the vineyards begin east of the city of Blois and continue past Tours to the west. Along the way, the river is joined by its tributaries the Vienne, the Indre and the Cher. The vines grow along these river valleys and on the plain of the Sologne between the rivers Cher and Loire

More than 4,000 hectares of vineyards grow within the Touraine AOC. They lie in the south-west of the Paris Basin, so the underlying geology is the sedimentary limestone found in all of the Touraine appellations. There are significant differences in the soils, however, which often go by local names, such as the flinty clay *perruches* and the warm and free-draining clay-limestone *aubuis*. There are also deep, sandy soils to the east in the Sologne, silty soils with or without stones, and alluvial gravel terraces. Many of the best terroirs are close to the banks of the Cher river and on the plateau between the Cher and the Loire.

At the beginning of the nineteenth century, Côt was the most cultivated grape variety around the banks of the Cher and it was only after the phylloxera crisis of the late 1800s that this situation changed, with the introduction of Sauvignon Blanc and Gamay. Today, Côt represents only 8 per cent of plantings and Sauvignon Blanc is by far the most important grape of Touraine. It accounts for more than 40 per cent of the vines grown. With rising prices of the more renowned Sauvignons of Sancerre and Pouilly Fumé further east, the cheaper Touraine Sauvignons, with their reliably fresh, fruity and uncomplicated flavours, have proved to be especially popular in the United Kingdom, which is the number one export market for these wines. These are early-drinking wines that typically have less flavour concentration and minerality than their more expensive counterparts in loftier appellations. They tend to be lighter in weight, often with attractive fruit that can range from nettles to passion fruit and white peaches, and slightly lower alcohol levels that can be a bonus in warmer vintages.

There is more to the Touraine appellation than Sauvignon Blanc, however. Gamay is also a major source of Touraine wines, accounting for around 20 per cent of plantings. The best of Touraine Gamay is light, fresh and juicy, offering straightforward enjoyment. It can be found in red or rosé blends, or made as a stand-alone variety, where it may be labelled 'Primeur' or 'Nouveau' if it is made by carbonic maceration, or 'Touraine Gamay' if vinified as a red. The latter must be 75 per cent Gamay, but the wine may also include a proportion of Cabernet Franc, Cabernet Sauvignon, Côt or Pinot Noir. This is where the appellation of Touraine starts to get a little more complicated. The different combinations and restrictions on which percentage of which variety can go into which blend are frankly baffling, and frequently changing.

The basic rules for Touraine whites are that they must be mostly Sauvignon Blanc, with up to 20 per cent of Sauvignon Gris (although see the DGC exceptions, below). Reds must comprise at least 40 per cent Côt and can be blended with Cabernet Franc, Cabernet Sauvignon, Gamay or Pinot Noir as winemakers wish. But due to the differences in climate in this broad appellation, if the vines grow west of the meridian of Tours, the wines must be made from at least 90 per cent Cabernet Franc instead. Touraine red wines are less uniformly reliable than the whites, perhaps because Sauvignon Blanc is a more straightforward wine with less variety in its vinification methods and style. However, the best reds offer medium-weight, plump, black- or red-fruited refreshment, usually destined for early consumption. Rosés are almost all dry and must contain two varieties, of which neither can comprise more than 70 per cent of the blend, and include Cabernets Franc and

Sauvignon, Côt, Gamay, Grolleau Gris, Pinot Gris, Pinot Noir, Pinot Meunier and Pineau d'Aunis. Sparkling wines produced in the traditional method make up a small proportion of production and may include all of the above varieties other than Gamay, plus Chardonnay, Orbois and Grolleau Noir. They age for nine months or more on lees before release.

Touraine – key facts

- **AOC granted**: 1939
- **Hectares in production**: 4,336
- **Annual production volume**: 200,000 hectolitres/26.7 million bottles
- **Styles**: 64 per cent white, 20 per cent red, 8 per cent rosé, 8 per cent sparkling
- **Grapes**: Sauvignon Blanc, Gamay, Côt, Cabernet Franc, Pineau d'Aunis
- **Soils**: very varied

Minor Touraine appellations

A further Touraine appellation (not a DGC) is Touraine Noble-Joué, a classification of an historic rosé grown in vineyards south of the city of Tours. This *vin gris* was hugely popular in the nineteenth century but slipped into obscurity as the conurbation of Tours encroached on the vineyards following the Second World War. It was eventually rediscovered by a group of vignerons in the 1970s, who applied for the remaining vineyards to be recognized with their own appellation, which was duly received in 2001. The small quantity of rosé wine made in this appellation is an aromatic, early-drinking wine with a distinctive *oeil de perdrix* colour.

Valençay is an historic appellation in the far south of the Touraine region, on sloping vineyards overlooking the Cher river. Evidence of winemaking here goes back to 965 and the three abbeys within the boundaries of the appellation suggest it was originally an important source of wine for the monks; latterly the wines were served at court at nearby Château de Valençay. Forests to the west make this a drier climate than is found in other vineyards in the vicinity. The majority of the production here is white from Sauvignon Blanc. Light and aromatic reds and rosés are made from blends of Gamay, Pinot Noir and Côt.

Valençay is the only place in France to have been granted an AOC of the same name for its wine and its cheese, an ash-covered goat's milk cheese shaped into a truncated pyramid. Naturally, it is a perfect match for the wines that share its name.

Historically an important wine-growing region for supplying the bistros of Paris, today Orléans is a shadow of its former self, with a mere 40 hectares of land planted. These vineyards are situated in the far east of Touraine at the most northerly point of the Loire's curving arc before it heads south towards Blois. This is a cool climate, and the wines are correspondingly light. Whites are from a majority of Chardonnay with Pinot Gris; rosés and reds are, unusually, made from a majority of Pinot Meunier, with the option of adding Pinot Noir or Pinot Gris. Orléans-Cléry is an even smaller appellation, solely for red wines from Cabernet Franc. The vines grow on ancient terraces of sandy gravel, yielding deeply coloured wines with a touch of spice on the finish.

Touraine Noble-Joué – key facts

- **AOC granted**: 2001
- **Hectares in production**: 34
- **Annual production volume**: 1,500 hectolitres/200,000 bottles
- **Styles**: 100 per cent rosé
- **Grapes**: Pinot Meunier (minimum 40 per cent), Pinot Gris (minimum 20 per cent), Pinot Noir (minimum 10 per cent)
- **Soils**: clay-limestone and flinty clay

Valençay – key facts

- **AOC granted**: 2004
- **Hectares in production**: 206
- **Annual production volume**: 100,000 hectolitres/1.3 million bottles
- **Styles**: 59 per cent white, 32 per cent red, 9 per cent rosé
- **Grapes**: whites – Sauvignon Blanc (minimum 70 per cent), Orbois Blanc, Sauvignon Gris, Chardonnay; reds – Gamay (30 to 60 per cent), Pinot Noir (minimum 10 per cent), Côt (minimum 10 per cent), Cabernet Franc (maximum 20 per cent); rosé – varieties as for the reds, plus Pineau d'Aunis
- **Soils**: Flinty clay *perruches* and sandy soils with a high proportion of stones

Orléans – key facts

- **AOC granted**: 2006
- **Hectares in production**: 40
- **Annual production volume**: 1,100 hectolitres/150,000 bottles
- **Styles**: 45 per cent red, 42 per cent white, 13 per cent rosé
- **Grapes**: whites – Chardonnay (minimum 60 per cent), Pinot Gris; reds – Pinot Meunier (70 to 90 per cent), Pinot Noir; rosés – Pinot Meunier (minimum 60 per cent), Pinot Gris, Pinot Noir
- **Soils**: sandy clay, sandy gravel and clay limestone

Orléans-Cléry – key facts

- **AOC granted**: 2006
- **Hectares in production**: 10
- **Annual production volume**: 330 hectolitres/44,000 bottles
- **Styles**: 100 per cent red
- **Grapes**: Cabernet Franc
- **Soils**: sandy gravel terraces

Touraine's Dénominations Géographiques Complémentaires

The DGCs of Touraine are not separate appellations in themselves, but can be appended to the name Touraine on the wine's label. They represent only 7 per cent of all the production of Touraine and are subject to more restrictive regulations regarding yields and varieties that can be used. For Touraine whites the basic permitted yield is 65 hectolitres per hectare, whereas for the DGCs it ranges between 52 and 60. For the Touraine appellation, reds and rosés can be produced at 60 hectolitres per hectare, but for the DGCs the figure is 55. Several of the DGCs also have minimum ageing requirements.

The first of these DGCs to be recognized was Azay-le-Rideau, whose vineyards border the valley of the Indre. This classification was introduced in 1953, followed shortly thereafter by Mesland and Amboise in 1955. Following phylloxera, most of the vineyards of Touraine had been replanted with Sauvignon Blanc and Gamay, but in these three areas growers were still cultivating the more traditional Chenin Blanc. The DGCs have since been updated to include reds and rosés.

From 1985 producers in the Cher valley began to plant Côt and Sauvignon Blanc in their best sites, while in the Sologne it became clear that it was the sandy soils that were most highly valued for Sauvignons. In 2011 a further two DGCs were recognized, those of Chenonceaux and Oisly. The Sauvignons of Touraine-Chenonceaux are notable for their consistent style, since the vines grow on slopes overlooking the river Cher and mature on lees, giving them added richness. The wines of Touraine-Oisly are noticeably riper in style, with higher alcohol levels than other Touraine Sauvignons and fruit profiles heading towards warmer-climate fruits such as apricots and nectarines, and even tropical mango notes.

Touraine-Azay-le-Rideau – key facts

- **Hectares in production**: 40
- **Annual production volume**: 1,000 hectolitres/130,000 bottles
- **Styles**: 67 per cent white (dry or *demi-sec*), 33 per cent rosé
- **Grapes**: whites from Chenin Blanc; rosés from majority Grolleau, plus Gamay, Côt, Cabernet Franc, Cabernet Sauvignon
- **Soils**: flinty clay *perruches* and clay-limestone *aubuis*
- Wines must be aged until 31 December of the year following the harvest.

Touraine-Amboise – key facts

- **Hectares in production**: 112
- **Annual production volume**: 3,500 hectolitres/500,000 bottles
- **Styles**: 47 per cent red, 43 per cent white (dry, *demi-sec* or sweet), 10 per cent rosé
- **Grapes**: whites from Chenin Blanc; reds from Côt; rosés from Côt (70 per cent or more) and Gamay
- **Soils**: flinty clay *perruches*
- Wines must be aged until 30 April of the year following harvest

Touraine-Mesland – key facts

- **Hectares in production**: 81
- **Annual production volume**: 2,900 hectolitres/390,000 bottles
- **Styles**: 55 per cent red, 36 per cent rosé (dry or *demi-sec*), 9 per cent white (dry or *demi-sec*)
- **Grapes**: whites from a majority of Chenin Blanc plus Sauvignon Blanc and Chardonnay; reds must be from 40 to 60 per cent Gamay plus Cabernet Franc and Côt; rosés from a majority of Gamay (minimum 80 per cent) plus Cabernet Franc and Côt.
- **Soils**: flinty, with miocene sands

Touraine-Chenonceaux – key facts

- **Hectares in production**: 161
- **Annual production volume**: 7,800 hectolitres/1 million bottles
- **Styles**: 80 per cent white, 20 per cent red
- **Grapes**: whites from Sauvignon Blanc; reds from Côt (at least 65 per cent) and Cabernet Franc (at least 20 per cent)
- **Soils**: homogeneous, on the higher slopes above the Cher river
- Whites must be aged until 30 April of the year following harvest, and reds to 31 August. The wines must have higher natural minimum alcohol levels compared to AOC Touraine and are subject to tasting.

Touraine-Oisly – key facts

- **Hectares in production**: 40
- **Annual production volume**: 1,800 hectolitres/240,000 bottles
- **Styles**: 100 per cent dry white
- **Grapes**: Sauvignon Blanc
- **Soils**: Sandy clay on the plateau between the Loire and Cher rivers
- Wines must be aged on fine lees until 30 April of the year following harvest and must have higher natural minimum alcohol levels compared to AOC Touraine. They must pass a tasting panel before release.

Notable Touraine producers

Domaine de la Rochette

Pouillé

www.vin-rochette-leclair.com

Domaine de la Rochette has been in the Leclair family for four generations and today it is in the hands of Vincent Leclair. He came back home to the domaine to help his father in 2005, turning his back on a career in mathematics. 'I missed the vines,' he says, simply. Vincent farms 50 hectares in Touraine and Touraine-Chenonceaux. Around half of the production is Sauvignon Blanc (the only permitted white grape). He also has the full mix of permitted varieties for reds and rosés.

In the vineyard, mechanization is used to the full, even to the extent of using a robot – the snappily named Vitibot – for working the soil in the vineyards. A recent purchase, this cost 200,000 euros, which seems like a lot until you consider that it was 50 per cent subsidized by the State, and that a tractor costs around 120,000 euros. And the cost of running the Vitibot is considerably lower. As Vincent points out, 'It's entirely autonomous; we take it into the vines, it works the soil and at the end of a row it turns around and carries on. It's electric, there's no need for a driver, it runs for 12 hours without a break and in the evening we bring it back to recharge it before taking it out the next day.' He's clearly delighted with it – and the fact that it costs a mere 10 euros a day to run (the price of 10 hours' overnight electricity). The harvest is also done mechanically, although there is no robot for that – yet.

For the whites, yields are kept to a reasonable level at 55 hectolitres per hectare for Touraine, and 50 for Touraine-Chenonceaux. When the grapes come in the whites and rosés go to ferment in temperature-controlled stainless steel 'for freshness and aromas,' Vincent says, and the reds ferment in concrete vats. After the reds are pressed, he does some micro-oxygenation on the press wine 'to oxidize it a little, bring out some of the chocolatey aromas, and to round the tannins.'

The wines are delicious representations of the simple pleasure of Touraine wines. The Touraine Sauvignon Blanc shows super fresh, crunchy fruit. Bottled young, this is crowd-pleasing Sauvignon at its best. The Touraine-Chenonceaux is less exuberant, but more complex and a little richer, with a silky, round texture. Vincent explains, 'The Touraine-Chenonceaux wines come from warmer, thinner soils that reflect the sunlight, and the vines have lower yields and riper grapes.' The wines with this DGC also need to be aged for 6 months before release. A rosé from Pineau d'Aunis (classified as IGP Val de Loire because it is 100 per cent of the variety instead of the blend required for appellation status) is delicate and juicy, and deliciously different. It has none of the peppery character that usually identifies a wine as Pineau d'Aunis. This is a character that Vincent considers a sign of underripeness in the grapes. A further red from Pineau d'Aunis also lacks the peppery notes, but is more red-fruited, a little Gamay-like, but with finely structured tannins. It is delightfully expressive, more so than the domaine's supple and easy-going Gamay, which is nevertheless highly drinkable. The Fleur de Printemps cuvée is an 80:20 blend of Gamay and Pinot Noir. It is a fragrant wine, with some earthy Pinot notes balanced with the exuberantly fruity Gamay. 'Pinot gives shape to the wine,' says Vincent. It's best served slightly cool.

It is good to learn that Domaine de la Rochette is in conversion to organic viticulture. These wines are all about immediate pleasure, and they do what they do very well.

Domaine Delobel

Ouchamps
www.domainedelobel.fr

Domaine Delobel winery is owned by young couple, Sandrine and Benjamin Delobel, both from the Touraine area. They bought their home, in Benjamin's village, in 2009 and managed to buy 14 hectares of vines in Touraine and Touraine-Oisly in 2014. The grapes in the latter area grow on a terroir of sand and clay that is classified only for Sauvignon Blanc. Both qualified oenologists, they make three cuvées of Sauvignon Blanc, including two Touraine wines, Angle Droit and Exponentielle, and one Touraine-Oisly, L'Ellipse. They also make 100 per cent Pineau d'Aunis wines: an aromatic rosé and a light, peppery red, both under the Vin de France label, and a Touraine Gamay.

One-third of the vines are more than 50 years old and Benjamin spends a lot of time taking care of his plants. He chooses to replace dying vines by layering (*marcottage*) to avoid having to use the clones he believes make the Sauvignon vines more susceptible to the trunk disease esca. At harvest time, almost all of the grapes are hand-picked and all are destemmed before a long, gentle pressing at the winery. Whites are fermented in tall, temperature-controlled stainless steel tanks that they somehow managed to install between the low wooden beams of their old *chai*. Benjamin controls the fermentation temperature for the whites at between 18 and 20°C, to bring out the 'exotic and citrus' character of the Oisly fruit.

Compared to the wines of Touraine, says Benjamin, the Oisly wines have 'more substance, they are more wines for gastronomy'. These wines are also required to have a higher minimum degree of alcohol and need to pass a strict tasting panel to be allowed to use the Oisly name.

Angle Droit is clean as a whistle, a fresh and citrussy Sauvignon with a grassy palate and zippy acidity. L'Ellipse comes from 50-year-old vines in Oisly. This is a ripe, full-bodied, interesting wine, with rounded fruit and substantial alcohol that is well integrated into the rich fruit. Exponentielle is vinified in 350-litre untoasted barrels, of which one-third are new. 'The idea is just to give structure,' explains Benjamin. This is another rich style, with intense but fresh fruit, and a vanilla roundness. The fruit is well supported by the oak, which does not dominate. In 2017 this wine was awarded the top prize at the world Sauvignon Blanc competition. Benjamin and Sandrine have, in a short space of time, succeeded in putting their young domaine at the front and centre of their region.

Domaine des Quatre Piliers

Noyers-sur-Cher
www.lesquatrepiliers.com

In the same way that some chefs can fashion the most humble of ingredients into a Michelin-starred meal, Valentin Desloges is a young man who has taken the ordinary and turned it into something extraordinary. In the lesser known reaches of the Touraine appellation, wines from the Cher have never enjoyed a stellar reputation, but Valentin is in the process of changing all this, sprinkling stardust on the unloved old vines of his homeland.

His father was a vigneron here, but Valentin took the opportunity to spend some time studying and gaining experience elsewhere before deciding on his path. 'I was lucky,' he says, 'my father's generation didn't have the chance to do that.' He studied in Tours and Bordeaux, before heading to Hunter's Wines in Marlborough, New Zealand, then returned to France to gain more experience at high-quality estates in Alsace, Saint-Emilion and Chassagne-Montrachet. Most recently he did a three-year stint at Coche-Dury in Meursault before deciding it was time to strike out on his own.

His vision was to make small-volume, high-quality wines, which are far from the norm in this part of the world. The local banks didn't share his belief that this was a sound business proposition, and he had to go further afield to find someone willing to take a risk on a young vigneron with a big plan.

His father's domaine was 30 hectares. Valentin kept just 4 of these, and managed to find some plots of old vines close by, bringing the new domaine to 10 hectares, with vines on either side of the Cher river. In all, he has 25 parcels, of which the largest is only 1 hectare. Half (5 hectares) of the vines are planted to Sauvignon Blanc, from which he makes two cuvées, one from younger vines and the other from vines that are 60 years old. Across the river, he owns 2.5 hectares of 70-year-old Cabernet Franc, situated high above the river in an open, exposed site on the left bank of the Cher. Here the soils are shallow red clay and limestone – 'hot and powerful soils – best for red wines,' says Valentin. The roots of the vines are quickly into the tuffeau bedrock that lies beneath the vineyards. He also has small parcels of Chenin Blanc, Pinot Noir, Côt and Pineau d'Aunis. Valentin declares no interest in blending these varieties, sticking to his mantra of 'one variety, one terroir,' saying, 'I just want to make a wine that has the taste of where it comes from.' And he is using his years of vineyard and winery training to do this to the very best of his ability. The vineyards are organic and the harvest is done by hand.

The winery, dug into the tuffeau, provides naturally cool temperatures year-round. Wood is used to age all of the wines, and for fermentation of the Sauvignons. Not just any wood, though: used wood from Bordeaux *cru classé* estates for the Cabernet Franc and Côt, and burgundy barrels for the Pinot Noir and Sauvignon Blanc. In the future, he plans to investigate using local wood from the forests where he likes to wander in his free time. The wines are for the most part unfined and unfiltered – 'We try not to cheat anywhere, and to respect each step' – and an enormous amount of tasting and testing takes place before these wines leave the cocoon of their barrel homes to be bottled. Conveniently, there is an oenology laboratory just a few kilometres from the winery, and Valentin is one of their most regular customers: 'We test every single barrel, and we have 130 barrels,' he tells me, adding, 'They must be tearing their hair out when they see us coming, because we often have 50 or 60 samples at a time for them to test.'

Tasting the wines in barrel feels like being led into a secret world of evolving creations that have never seen the light of day. A Pineau

A voyage of discovery: tasting barrel samples with Valentin Desloges

d'Aunis is fragrant with raspberries and white pepper, a young Cabernet Franc is the epitome of delicacy, his Côt is spicy and brooding, with black plums and grilled meat character. When will it go into bottle, I wonder? 'When it's ready.' The finished wines are no less impressive. A Pinot Noir (long since sold out) showed fabulous aromatics and effortless elegance. The vines that produced the wine are only 20 years old: this is a future show-stopper. His '*grande cuvée*' Cabernet Franc, Au Dessus de Vitré, is beautifully perfumed and juicy, with rich black fruit. It is perfectly judged, with tannins that melt in the mouth. The Sauvignon Blancs are a far cry from the simple refreshment offered by most wines in the appellation. The Bel Air in particular, from 60-year-old vines, recalls a refined, featherlight Sancerre, with its chalky fruit texture and lightly creamy, lees-aged character.

With only a few vintages under his belt (the first was 2020), Valentin Desloges is already one of the new stars of Touraine, and indeed all of the Loire Valley. Self-effacing to a fault, he says, 'I've realized there is nothing new to invent. Everything is here; our work is simply to reveal it.'

Domaine Mérieau

Montrichard

www.merieau.com

Jean-François Mérieau is a busy man, which is not surprising considering he is running a 40-hectare domain, producing more than a dozen wines, and has an active tourism business in the popular tourist town of Montrichard on the Route Touristique de la Vallée du Cher. His wife, Emilie, works alongside him managing tourism, communications and much else besides.

The domaine was created in the 1920s by Jean-François' great-grandfather, in the times when wine was part of a polyculture. It passed from father to son in 2000 and Jean-François has worked hard to improve every aspect of the business. He says he has worked for the last ten years to manage frost damage by trying to slow down the vine through judicious pruning. 'We are always trying to adapt to the climate,' he says.

The main winery and vines are located in St-Julien-de-Chédon, just south of the Cher river. The vineyards are planted 50 per cent each with red and white varieties, although the domaine has more red wine cuvées than white. The reds are all harvested by hand and for the whites it is a mixture of manual and mechanical harvesting. Over the years the domaine has become progressively more attuned to nature. When Jean-François took over from his father, he stopped using herbicide, and the domaine has been certified organic since the 2023 vintage.

A range of characterful wines is produced by this domaine, with equally characterful, colourful labels and names that must have generated a lot of fun in the creation, as they are all plays on words. A traditional method sparkling comes in an especially festive package of colourful polka dots on the label (or perhaps they are balloons? Either way, it's tempting to pop the cork). The wine, J'Ose les Fines Bulles is a joyful blend of Chenin, Menu Pineau and Chardonnay that spends two years on lees and a further year ageing at the domaine. It has yeasty breadth and masses of appeal. The Sauvignons are clean and correct, ranging from green and zesty to juicy and exotic. There is a lovely, light Gamay, Le Bois Jacou, but the most interesting of the reds for me is the Cent visages Côt, which comes from the grapes of 50-year-old vines that ferment and age in concrete. Unlike some of the domaine's wines which are aged in oak, this has startling clarity of bright fruit and is deliciously pure.

The domaine is happy to receive visitors to its well-stocked tasting room in Montrichard, where there is also a restaurant in the summer months, as well as organized scooter rides and meals in the vines for groups.

La Grange Tiphaine

Amboise

www.lagrangetiphaine.com

Coralie and Damien Delecheneau run this quality and naturally-focused domaine in the outskirts of one of the most visited towns of the Loire Valley. They met while studying for their oenology diplomas in Bordeaux and subsequently went in separate directions, Coralie to California and Damien to Chablis (Laroche) and Bordeaux (Lafon-Rochet) before working together in South Africa.

When they returned to Damien's home in Amboise in 2002, Damien took over his family's wine estate, with a major overhaul towards producing quality wine. Starting with a new name for the domaine to mark a new beginning, they changed just about everything, converting to organic viticulture by 2011, and subsequently certifying the estate, now all harvested by hand, to biodynamic viticulture. One of the features of the domaine is its high percentage of old vines, which contributes to the quality and finesse of the wines.

They farm 17 hectares in the appellations of Montlouis-sur-Loire, Touraine and Touraine-Amboise. In Montlouis they produce six different wines at all sweetness levels, including a delicious single-fermentation *pétillant originel*, Nouveau Nez and the mouthwateringly mineral Clef du Sol, the latter from 80-year-old vines.

In the Touraine-Amboise appellation, the Bel Air cuvée comes from vines aged 7 to 77 years old. It is an uncomplicated, leesy style with good concentration of peachy fruit. The red Touraine-Amboise wine is the Côt Vieilles Vignes. '*Vieille*' can mean almost any age on a wine label, but here it is an understatement, for these vines are more than 100 years old. This gives the grapes layers of flavour and a natural finesse. The wine is an inky, spicy mouthful, with beautiful natural freshness and very supple tannins. The aim here is to manipulate as little as possible, says Damien, so there is little in the way of tannin extraction. The grapes are macerated and fermented in concrete with indigenous yeasts, and the wine spends 9 months in 225-litre barriques. It's a lovely wine, unlike anything else.

This is a domaine where one can buy with confidence across the range.

Lionel Gosseaume

Choussy

www.lionelgosseaume.fr

Lionel Gosseaume is from a winemaking family; his father started his own domaine in 1964. But Lionel wanted to carve his own path. Rejecting involvement in the family business as a young man, he worked instead in various alternative aspects of the wine business, helping young wine growers, and subsequently representing them all over France. After 19 years, he decided to return to his roots and start up on his own. He took out a bank loan, rented some vines and built a functional, compact winery.

His first vintage was 2007, and he has successfully built the business to a point where he now farms 35 hectares and exports around half of his wines, which is more than double the average exports for the Loire as a whole. His previous career working with viticulturists has served him well in his new life as a vigneron. Recognized by his peers, he was until recently the president of the growers' association for InterLoire, the professional body representing all Loire growers in the western Loire, and he is still very involved in representing growers locally.

Sauvignon Blanc accounts for 28 hectares of his domaine but, Lionel says, 'I don't sell Sauvignon Blanc; I call it Touraine.' For his main cuvée, Les Sauterelles, he says he is looking for 'very intense aromatics, citrus, grapefruit, a vegetal zestiness, and not too much maturity.' The wine delivers a mouthful of great-value Sauvignon, with aromatic intensity, crunchy fruit and wonderful freshness. 'It should be open and upright, with volume and length,' says Lionel, 'for drinking as an *apéro* or with food.' It's really very easy to drink. The secret to the quality of these Touraine Sauvignons, says Lionel, is the terroir: 'We have deep clay soils with good water reserves. To achieve good aromatics and acid tension, the vine mustn't suffer.' He also makes a Sauvignon in Touraine-Oisly, a warmer terroir due to its position on free-draining, sloping, sandy soils. The wines typically have a degree more alcohol than Touraine wines. Lionel's Touraine-Oisly Climat No.1 is a powerful and quite weighty

Sauvignon, with very ripe fruit and a lingering intensity on the palate. It spends a year ageing in tank before release.

Another white, Climat No.2 is a 50:50 blend of two ancient and little-known varieties, Meslier Saint-François and Menu Pineau (also known as Orbois). This wine spends two years ageing in a tank with no protection against oxidation other than leftover lees from Lionel's Sauvignon Blanc fermentations. It's a fascinating wine, with a deep colour, creamy and oxidative aromas and a rich, savoury character – definitely a food wine.

Lionel also has a few parcels of Gamay, Côt, Pineau d'Aunis, Cabernet Franc, Chenin and Chardonnay, from which he creates a colourful range of wines. Les Galipettes is an inexpensive dry rosé made from Pineau d'Aunis, Gamay and Côt. It is aromatic, vinous and fruity, and beautifully presented – it's another great-value wine. Les Marcottes is a Touraine Gamay that comes from old vines, harvested by hand. The just-ripe, crisp fruit is bright and spicy. Half of the wine is made in the carbonic maceration style, and half as a regular red wine with pump-overs. The aim, says Lionel, is not to extract tannins. It is savoury and interesting. Climat No.3 comes from a massal selection of old Côt vines from warm soils with a high percentage of flint. This is a low-volume, hand-made wine that is not made in every vintage. It has aromas of pepper and very ripe, confit black fruit. It is gulpable, soft and juicy with savoury notes of blood and iron filings. Full and expressive, it's a lovely example of what can be achieved with this variety in Touraine, something that can be said, in fact, about most of the wines of Lionel Gosseaume.

Other producers to try

Domaine Jacky Marteau, Joël Delaunay, Vignoble Dinocheau, Domaine de la Renaudie, Château de Fontenay, Domaine Octavie, La Cave du Père Auguste, La Piffaudière (Olivier Bellanger), Domaine des Echardières, Jean-Christophe Mandard, Domaine de l'Azuré, Seneau & Fils, Domaine des Corbillières, Vins Beaujardin, Vincent Roussely, Cocteaux Amélie & Franck.

7

CENTRE LOIRE: LAND OF SAUVIGNON BLANC

The Centre Loire vineyards are located in the heart of the country, not far from the city of Bourges. This is also the centre of the Loire itself; the bridge across the river to the village of Pouilly marks the halfway point of the Loire's journey to the Atlantic Ocean. The river flows close to the major appellations of Sancerre and Pouilly Fumé, which face each other across the water, and the vines of the Coteaux du Giennois grow on the right bank of the river north of Pouilly. Menetou-Salon is a natural extension of the Sancerre vineyard area, continuing to the south-west. The vineyards of Reuilly and Quincy are clustered around the Cher and Arnon tributaries of the Loire, a little further south. Châteaumeillant is a little-planted and little-known red wine appellation at the edge of the Massif Central.

White wines make up 86 per cent of production. These vineyards are dominated, both in quality and quantity, by the Sauvignon Blanc grape. Other than the tiny Chasselas appellation of Pouilly-sur-Loire, Sauvignon is the sole permitted variety for white wines in any of the appellations of the Centre Loire. Perfectly suited to the limestone-clay and flint soils of the Sancerre and Pouilly Fumé hills and valleys, the grape achieves an intensity of mineral expression found nowhere else, and the top growers here are making truly exciting, terroir-focused wines. Sancerre is the appellation driving the quality of the wines, which has surged in recent years. It is the most highly demanded, most-exported and best-known of the Loire Valley's AOCs. Pouilly Fumé, slightly less successful than its larger neighbour, is nevertheless a source of similarly sublime Sauvignon when it is in the right hands. Less obviously mineral but still appetizing

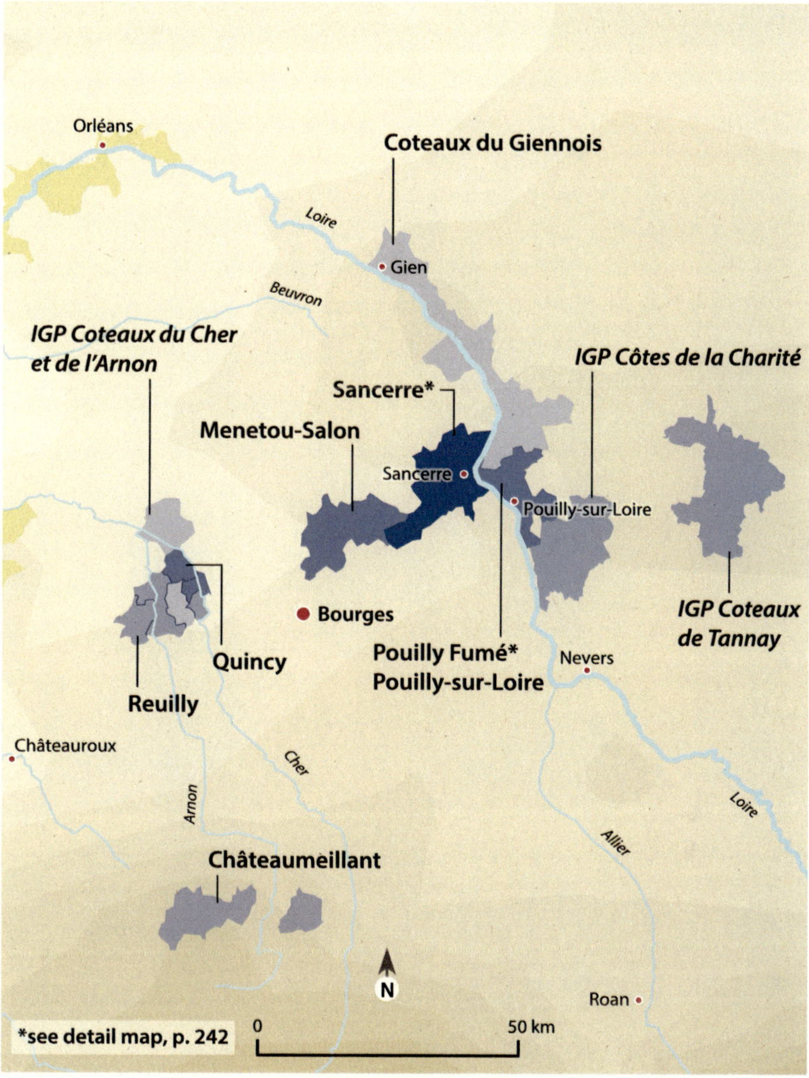

Map 9: Centre Loire

expressions of the grape can be found in the satellite appellations of the Centre which, with less renown and lower prices, can offer good-value alternatives to Sancerre and Pouilly Fumé. In most of the appellations of the Centre, yields for Sauvignon can be as high as 65 hectolitres per hectare, but the better growers restrict yields to well below this level.

The only other grape of significance in the Centre Loire is Pinot Noir, which accounts for 15 per cent of plantings. In Sancerre, Menetou-Salon and Reuilly it is vinified alone, and the top Pinot Noirs from Sancerre are the most serious French wines from the grape outside Burgundy. In the Coteaux du Giennois and Châteaumeillant Pinot is blended with Gamay for early-drinking, soft and fruity reds that retain the Loire's characteristic lifted acidity. In Reuilly there are some attention-worthy *vins gris* made principally from Pinot Gris.

To the south and east respectively of the AOCs on the the right bank of the Loire are two IGPs, Côtes de la Charité and Coteaux de Tannay, where reds, whites and rosés can be found and Chardonnay replaces Sauvignon as the white grape of choice.

Centre Loire – key facts

- **Hectares in production**: 6,000 (11 per cent of Loire total)
- **Annual production volume**: 330,000 hectolitres/44 million bottles (12 per cent of Loire output)
- **Styles**: 86 per cent white, 10 per cent red, 4 per cent rosé

Centre Loire appellations

- Sancerre
- Pouilly Fumé
- Pouilly-sur-Loire
- Coteaux du Giennois
- Menetou-Salon
- Quincy
- Reuilly
- Châteaumeillant

SANCERRE, POUILLY FUMÉ AND POUILLY-SUR-LOIRE

Sancerre: easy to say, easy to understand, easy to drink and, it seems, easy to sell. These crisp and vibrant white wines are the Loire's most reliable calling card worldwide and by far its most successful export, with up to 70 per cent of the wines sold outside France. These fresh, zesty, mineral-infused Sauvignons are immediately enjoyable, and the best can age for a decade or more. This is one of the few Loire appellations where the presence of the vine is ubiquitous. An area of 3,000 hectares is devoted to just two grape varieties: Sauvignon Blanc and Pinot Noir.

Sauvignon Blanc accounts for more than 80 per cent of Sancerre's production. The other famous Sauvignon of the Centre Loire is Pouilly Fumé, also known as Blanc Fumé de Pouilly, an appellation based on the right (east) bank of the Loire around the town of Pouilly. Pouilly Fumé wines are in the same style as Sancerre, with a little more breadth and sometimes smoky, gunflint aromas. (This is one proposed explanation for the name 'Fumé', meaning 'smoked', the suggestion being that it comes from the flinty soils in parts of the appellation. Another is that the term originally derived from the greyish bloom visible on the surface of the grapes at harvest time.) The second white wine appellation in the same area is Pouilly-sur-Loire. Rarely seen and barely planted, this is an appellation solely for Chasselas. Formerly successfully grown as a table grape, when yields are managed it can produce pleasantly delicate wines with a light, floral scent.

Sancerre is one of the prettiest places to visit in the Loire Valley and, indeed, the country. Voted France's favourite village in 2021, this charmingly authentic medieval town captures the spirit of times past, with its slightly dilapidated, wood-shuttered stone houses, out-of-control climbing plants, mossy, red-tiled roofs and narrow, cobbled streets that bear such evocative names as 'Boulevard de la Soif' and 'Rue des Pressoirs'. And there is a castle on top of the hill, all the more alluring for being privately owned and not open to the public. Its dramatic hilltop position means there are no bad views in Sancerre. To the east, the land falls away steeply to reveal the Loire, snaking its way

Sancerre viewed from the vineyards of Pouilly Fumé

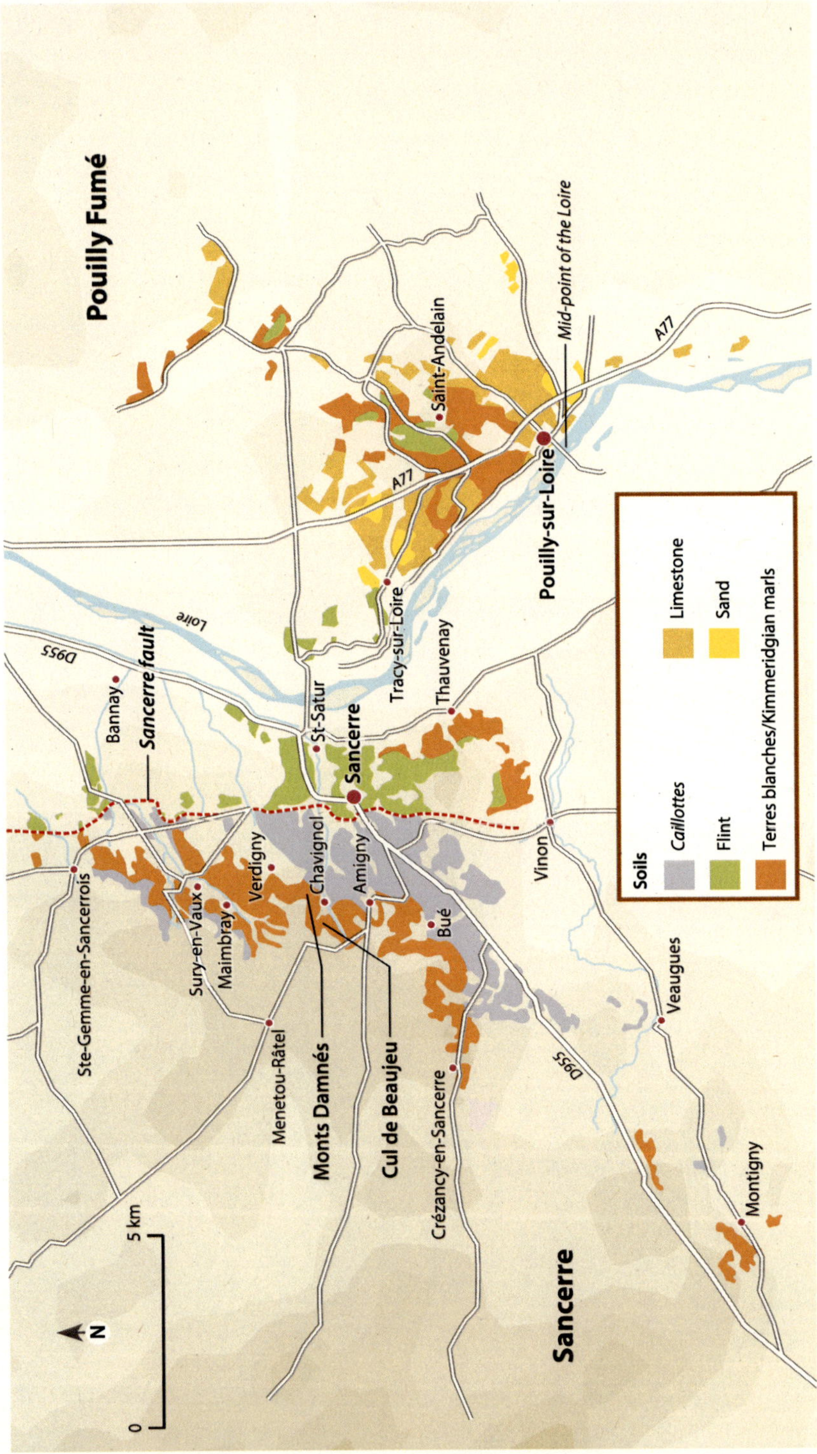

Map 10: Sancerre and Pouilly Fumé

northwards, and the vineyards of Pouilly Fumé beyond. To the west, north and south, a succession of green, rolling hills and the half-hidden wine villages of Sancerre spread as far as the horizon.

Close to 2,500 hectares of Sauvignon Blanc criss-cross the landscape surrounding Sancerre, and there are a further 1,400 hectares in the Pouilly Fumé appellation, creating an undulating ocean of vines in all directions. The Sancerre vineyard is one of the few in the Loire to have pronounced slopes, and these are often the source of the best wines: the Monts Damnés and Cul de Beaujeu are probably the best-known of these, both rising above the village of Chavignol to a height of nearly 400 metres. Only the vineyards of the Auvergne to the south can boast higher altitudes in the Loire. Pouilly, in contrast, is less markedly hilly and the altitude lower. From the cosy, village feel of Sancerre, crossing the river can come as a bit of a shock. In place of the charm of the country roads that meander between the peaceful villages of Sancerre, the traffic of the A77 thunders past the vineyards of Pouilly Fumé. This major route to the south was formerly an important source of passing trade for the local vignerons.

But it's not just the river (or the road) that separates these neighbouring appellations. Despite their proximity both geographically and in terms of their wine styles, the two appellations feel very separate. Visit

Chavignol and Sancerre from the Cul de Beaujeu vineyard

a restaurant in Sancerre and you'll be unlikely to see any Pouilly Fumé on the wine list. Pouilly vigneron Alexandre Bain says, 'Pouilly Fumé lives in the shadow of Sancerre.' He's not just talking about the fact that Sancerre is larger and undeniably prettier. These days, Sancerre is simply the more successful of the two appellations – and if Sancerre's fortunes are good, they raise the prices of land, grapes and wines of its neighbour across the river. While this might be true for most producers, the best of the wines of Pouilly Fumé (including those of Alexandre Bain) are not inexpensive and certainly rival the best of Sancerre for quality. It was in Pouilly Fumé that the late Didier Dagueneau revived the notion of quality in the 1980s, most notably with his cuvée named Silex (meaning flint), which has since spawned multiple imitations from vignerons wishing to highlight their own flinty terroirs.

Sauvignon is perfectly suited to the slopes and soils of Sancerre and Pouilly Fumé. But in the same way nobody speaks of Chardonnay in Burgundy, the winemakers here rarely mention the grape, and some even profess a dislike of its character. In short, nobody wants their wine to taste of Sauvignon in the form it is best known elsewhere. If a wine has the explosive and pungent aromas of gooseberries, grass and cats' pee, it's broadly considered a failure, for the highest aim here is terroir expression. And it is terroir that is the calling card of the great wines of

Kimmeridgian marl is packed with fossilized marine life

Sancerre and Pouilly Fumé. Talented vignerons are taking an increasingly *parcellaire* approach, identifying the particularities of individual sites and capturing these essences of place in a bottle.

The Loire's most famous Sauvignon Blanc appellations have a range of similar soils. These are mostly clay-limestone and Kimmeridgian marls, with a minority of flinty clay. The younger, flinty soils are found around the centre of Sancerre and to the east around Saint-Andelain in the Pouilly Fumé appellation. The soils in the more westerly parts of Sancerre are the older limestone-based sedimentary soils of the Paris Basin.

Understanding the soils of Sancerre

Sancerre has been described as a unique geological site, a *mille-feuille* of terroirs that dates back to the Jurassic, Upper Oxfordian and Kimmeridgian periods, between 150 and 160 million years ago. Like many other parts of the Loire Valley (and the vineyards of Champagne and Burgundy), Sancerre lies within the ancient former sea of the Paris Basin. What distinguishes the geology of Sancerre, however, is the Sancerre fault (also known as the magnetic anomaly of the Paris Basin). The movement of this major fault, along with associated more minor faults close by, led to the folds we see in the landscape today and the complex combinations of terroirs, often from different eras, existing side by side.

A few metres separate 100 million years of history along the Sancerre fault

The Sancerre fault runs north to south through the region, straight through the town of Sancerre. In fact, just south of Sancerre, a path between the vines follows the line of the fault so clearly that it is easy to see the differences in geological eras in the vineyards either side of the path: on the one side, cool, rounded limestone *caillottes*, on the other flints, warm to the touch.

The flints are the youngest soils, at 40 million years old. These make up only a small proportion of the Sancerre vineyard, around 15 per cent, and a similarly small proportion in Pouilly Fumé. These smooth, sharp, impenetrable stones can be seen at the surface of the vineyard; they retain and reflect the sun's heat. This warming effect, and the free-draining nature of these soils compared to clay and limestone, together create a drier and more extreme growing environment for the vines. Wines from grapes grown on flint typically show power and mineral intensity and have good ageing potential.

A few kilometres west of the town of Sancerre are the oldest soils, stony clay-limestone known as *griottes* or *caillottes,* depending on the size of the stones (*griottes* are smaller). These rocky soils make up 40 per cent of the vineyard and are around 150 million years old. They give early-maturing, fruity wines with fine acidity. In the west of the appellation are the slopes of Kimmeridgian marls. Mostly clay with around 30 per cent limestone, they are known as *terres blanches* because the rich profusion of marine fossils gives the soils a whitish appearance. These soils are found in 40 per cent of the vineyard and are around 140 million years old. Any handful of soil from the prized Kimmeridgian marl vineyards of Monts Damnés or Cul de Beaujeu vineyards is likely to yield a mixture of tiny oyster shells. Larger fossilized molluscs are also easily discernible, scattered all over the surface. These soils have the same origins as those of Chablis, less than two hours' drive to the east.

Pinot Noir is a minority grape in Sancerre. This is perhaps the only region in the world where the revered Burgundian variety is considered a minority interest, often consigned to poorer sites throughout the appellation that are less suitable for Sauvignon Blanc. This has not always been the case, however. Until phylloxera struck, in 1885, Pinot Noir vines were in the majority in most parts of Sancerre, though seemingly not in Chavignol, where Sauvignon has always held sway. The dictates of fashion – and possibly ease of productivity – led producers to experiment with Sauvignon when rebuilding their vineyards after phylloxera. The combination of Sauvignon and the chalky soils of Sancerre proved to be an instant hit, both locally and in Paris, where much of the wine was shipped. Nonetheless, Sancerre reds are the finest examples of Pinot Noir in the Loire. The best combine perfume, silky fruit and always-bright acidity. Rosés from Pinot Noir comprise a minority of production, but are often of high quality, with correspondingly high price tags.

Sancerre – key facts

- **AOC granted**: 1936 for whites, 1959 for reds and rosés
- **Hectares in production**: 3,000
- **Annual production volume**: 174,000 hectolitres/23.2 million bottles
- **Styles**: 84 per cent white, 10 per cent red, 6 per cent rosé
- **Grapes**: Sauvignon Blanc, Pinot Noir
- **Soils**: Kimmeridgian marls – (clay-limestone *terres blanches*), limestone *caillottes* and *griottes*, and flinty clay.
- **Villages**: 14 – Banny, Bué, Crézancy, Menetou-Ratel, Ménétréol, Montigny, St-Satur, Ste-Gemme, Sancerre, Sury-en-Vaux, Thauvenay, Veaugues, Verdigny and Vinon

Notable Sancerre producers

Alphonse Mellot

Sancerre

www.mellot.com

For someone whose family history in Sancerre dates back to the sixteenth century, Alphonse Mellot seems to be in an awful hurry. When I visited he was dressed in running gear and by the end of our visit, I wished I was too. We whizzed through the vineyards and then speed-tasted our way through the Tardis-like network of ancient cellars at the winery, twisting and turning beneath the streets of Sancerre. Along the way we passed steel and concrete tanks, wooden vats, an oak 'egg' and an extraordinary rotating oak sphere.

Alphonse is the nineteenth generation to bear the family name. To make his mark right from his first vintage (1990), he claims, 'The first thing you have to do is kill your father.' In other words, he needed to be different. Speaking of his red winemaking, he explains, 'At that time, I wanted to show myself [he puffs out his chest for emphasis] – but I was extracting too much and using too much oak. I am incapable of drinking those wines now – a wine with too much oak is a wine that is stillborn.' After a 2009 'epiphany' when tasting a 2002 Musigny from De Vogüé in Burgundy, he says he changed everything.

All of the domaine's grapes have been certified biodynamic since 1999, and all are farmed biodynamically and hand-picked. Most of the wines come from a large block less than a kilometre south of Sancerre, called La Moussière. The 38-hectare calcareous site is the source of the domaine's largest cuvée of the same name, as well as the smaller Edmond and elite Génération XIX wines, sourced from very old vines. He also has plantings in other terroirs close by, where he makes eponymous cuvées Le Paradis, Les Romains and Les Demoiselles. Appellation regulations stipulate that Sancerre plantings must be over 6,100 plants per hectare, but Alphonse prizes density and has planted 20,000 Pinot Noir vines per hectare for his wooden sphere-fermented rosé, Vingt Mille Pieds Sous Sancerre, an irresistible, full-bodied and lushly-textured wine that is, he assures me, 'the best rosé you'll ever taste in your life.'

The nineteenth Alphonse Mellot

His production is two-thirds white and one-third red and rosé. There is huge attention to detail at every stage of production. Wines are marked by rich, concentrated, ripe fruit, a strong personality, and a sense of place.

The domaine's wines are exported worldwide and there is a dedicated shop in the centre of Sancerre that showcases the range. Alphonse is also a champion of nearby IGP Côtes de la Charité, where he makes impressive Chardonnay and Pinot Noir under the brand name Les Pénitents, demonstrating the potential of this little-visited, rural corner of the Loire.

Claude Riffault

Sury-en-Vaux

www.clauderiffault.com

Before you've even tasted a drop of wine, everything about this domaine smacks of quality – from the cool, metal-caged stone walls to the well-organized cellar and the sparkling Zalto tasting glasses. This is the work of Stéphane Riffault, son of Claude, who is now in charge of the family estate. He works alongside his Burgundian wife, Bénédicte. The business of wine growing was started by Stéphane's great-grandfather in the nineteenth century but, over the years, wine grew in importance as the focus of income for the family. Stéphane trained and worked in Burgundy, and it is this experience that has shaped the direction and quality of the domaine since he took over from his father. As with most of the top growers, Stéphane is searching for individual terroir expression in his wines. He farms 60 plots in and around Sury-en-Vaux, making a top-class range of red and white Sancerres.

The domaine has been certified organic since 2016, and biodynamic since 2021. There are no chemicals used in the vineyard and everything is harvested by hand. In the winery, Stéphane uses only native yeasts, a choice which means, he says, 'the fermentations can take longer to finish – sometimes up until December.' To give himself as much flexibility as possible during the harvest, he has invested in additional presses and now has four to cover the 15 hectares of annual crop. Fermentations take place in wood of various sizes, from 350-litre barrels to 20-hectolitre tanks. The size of the barrel is important, but even the thickness of the staves plays a part. 'I use thicker barrels for wines that need longer ageing; for those to be released earlier, I use thinner staves to increase oxidation,' says Stéphane. For the reds he uses whole-bunch

Stéphane Riffault: a master of the terroirs of Sancerre

fermentations with no crushing or punching down (*pigeage*). Whites are aged on total lees with some *bâtonnage* – 'I know it's not fashionable, but I do a little, depending on the cuvée,' he says, 'It adds complexity.'

The wines are strikingly good, showing crystal-clear, precise fruit and a real sense of place. It takes confidence and knowledge to produce wines like this, and in this domaine both are evident in spades. Tasting wines with Stéphane feels like taking a walk up, down and around the slopes of Sancerre to discern the smallest differences between the plots that he knows so well.

He has land on all the main terroirs of Sancerre. Grapes growing on flint go into the cuvée Les Chailloux, an upright style of mineral purity and salinity. It's wonderful. Les Chasseignes is a predominantly limestone soil (*caillottes*), giving a wine of delicate smokiness and fine, sweet fruit. Les Denisottes comes from older vines grown on soils with a higher percentage of clay. It is a firm wine of energy and personality, with a bitter minerality on the finish. Two *monoparcelle* wines grow on adjacent, south-facing plots of Kimmeridgian marl. Monoparcelle 538 is the higher and windier of the two sites, with thin topsoil. It gives a superbly elegant, floral, delicate, juicy wine. Monoparcelle 469 is slightly lower in altitude, with slightly deeper soil. It has less wind and more temperature variation between morning and afternoon. The difference between the two sites shows the extraordinary variations that can result from these subtleties in terroir. The 469 has rounder fruit and a little more weight than 538, backed by a mineral intensity and savoury bitterness to the tangy fruit. It is less delicate, but no less impressive. Hard to find, Stéphane's La Noue is among the best reds of Sancerre, a most wonderfully scented and succulent Pinot with silky texture and perfect balance.

Domaine Daniel Chotard

Reigny

www.chotard-sancerre.com

Nestled at the foot of an amphitheatre of vines is the hamlet of Reigny, near Crézancy, and here is the home of the Chotard family. Daniel and his wife Brigitte built the reputation of the domaine in the 1990s and live practically within the walls of the winery. But they have now handed over to their son, Simon, who runs the estate with his wife, Juliette. 'We did the 2011 and 2012 vintages together and then he left me to it,' says Simon, who has also spent some time working at Allan Scott in Marlborough, New Zealand, checking out the competition from down under.

Simon and Juliette met in their biology class in 2007. He went on to study oenology and she continued to become a biochemist, so it feels like the estate is in safe hands. Under the young couple's management, the domaine's 17.5 hectares are now certified organic and there is an increased focus on cuvées that reflect the varying soil types of Sancerre. Their basic white is a blend of each of the three terroirs. It is a ripe, peachy, altogether welcoming sort of wine, with lovely balance and concentration. Like all the Chotard wines, it has no added yeasts, and spends some time ageing on lees. The wines might ferment in stainless steel, oak foudres, barrels or amphorae, depending on the desired style.

The oldest vines of the domaine are up to 80 years old and the fruit from these is destined for Marcel Henri, a delicately floral wine with minerals, smokiness, and a long, linear, salty finish. Juliette describes

Les Cris as her favourite plot, due to the mineral character it imparts to the wine. 'It has higher levels of clay, which keeps freshness,' adds Simon. The Les Cris wine has maturity and ripeness without any trace of heaviness. Mineral and intense, it shows a bitter twist of citrus peel and savoury length. Les Coutones is a steep, south-facing slope that retains freshness due to the clay-rich soils that inspired the muddy boots on its label (the message: clay soils are sticky). The wine from here is complex and broad, with mouthwatering freshness and bucketfuls of charm. From a south-east facing slope of dark Kimmeridgian marls comes Racines, a firm and deliciously mineral expression of place. 'This has less tension and more density,' explains Simon.

The reds are a minority of production but their 4 hectares of Pinot Noir produce some of the better examples to be found in Sancerre. Les Grandes Roulottes is soft, juicy and easy, while Les Champs d'Amigny is a more serious mouthful, with powdery tannins and spice, and smoky, herbal fruit. A five-year-old La Vallée de la Côte is supple and gently evolving. It had a month of maceration, Simon tells me. 'We work slowly,' he says, 'trying to get just the right level of extraction.' Le Chant de l'Archet comes from the oldest vines of the domaine. It shows a savoury Pinot Noir character, with juiciness and a depth common to all the wines that come from this high-quality estate.

Domaine Denizot

Verdigny
www.domaine-denizot.com

Thibauld Denizot is the sixth generation of his family to farm grapes in Amigny and the neighbouring villages of Verdigny, Chavignol and Sancerre. But he is the first to bring a focus on quality and bottled wines. When he arrived at the domaine in 2005, everything was sold in bulk. Since taking over from his father in 2016, he has changed much of the work in the 45 parcels and 20 hectares he farms. Vines are carefully pruned – twice – to minimize damage from the vine disease esca to which Sauvignon Blanc is especially prone. He has also eliminated the use of chemicals in the vineyard, with plans to be fully organic from 2024.

The aim with the wines, explains his wife Jennifer, is 'not to fall into the varietal character.' This is the goal of many a quality-minded Sancerre producer and is achieved, she says, by harvesting later and ensuring yields are controlled. 'The limestone soils already give plenty of acidity,' says Jennifer, 'so we can take our time with the harvest.' The basic Sancerre comes from multiple plots and is aged for a year on fine lees before release. Unoaked, it is bright, mineral, citrussy, clean and fresh, fragrant and delicious. The site-specific wines have up to 18 months' ageing. They ferment in untoasted 600-litre *demi-muids* and 350-litre *cigares*, constructed using water vapour to avoid undue wood influence in the wines. Les Bouffants comes from a warm site of *griottes* soil. The wine combines weight and richness with a surprisingly vertical acidity. Harmonious and bright, with lemon and pineapple fruit, a touch of smoke lingers on the finish. Damoclès is named for the steeply sloping site, Côte de l'Epée, 'slope of the sword', a sunny, south-east-facing site with very thin limestone *caillottes* topsoil. It's a striking wine, mouthwateringly saline, slicingly acidic but not angular, with beautiful concentration and fluidity. From the domaine's two hectares across the Loire come two Pouilly Fumé wines, including the delicious Séléné, a round and sweetly fruity wine with good minerality. Named after the goddess of the moon, the label depicts the idyllic view of the sky from the couple's terrace at their home in Sancerre, looking towards the vineyards of Pouilly. There is also a rosé and a tiny amount of red made at the domaine.

This is a lovely range of wines. The couple has joined with some of the other top growers in opening an independent wine shop, Vignerons La Cave (see p. 273), in the centre of Sancerre, selling the best wines of the region and a good selection from elsewhere in France.

Domaine Edmond Vatan/Clos la Néore

Chavignol

Anne Vatan took over the family vineyards from her father Edmond in 2008. She divides her time between two lives, as a professional historian and archaeologist in Paris, and producer of a single and almost unobtainable wine, the Clos La Néore in Chavignol. The vineyard is a single hectare, situated in the most covetable section of the Monts Damnés vineyard. It is planted with old vines that came from a massal selection planted by her father in the 1970s, when the domaine was considerably larger. When Anne took over, she decided to focus solely on this jewel of a site. Viticulture is organic and yields low. The winemaking is simple: the must is fermented in four temperature-controlled, 1970s enamel-lined vats and kept in bottle for three years before release. From 2018, Anne bottles only in magnum because, she says, she wants to encourage people to keep her wine and a magnum is less likely to be opened too early. Clos La Néore is taut and mineral in its youth, but this aristocratic wine is well worth the wait – if you can find it.

Domaine Fouassier

Sancerre

www.fouassier.fr

Cousins Benoît and Paul are the current descendants of a long line of Fouassiers that goes back ten generations. They recount that the current strength of the estate is largely down to their enterprising grandfather. He replanted the first vines after phylloxera, in 1902. Working alone from the age of 18, with ailing parents, he built the family business back up by exchanging plots with neighbours who did not want the trouble of starting over. Today the domaine comprises 60 hectares in Sancerre, producing 20 separate cuvées to represent each terroir. The Fouassiers have been doing this since the 1980s, long before it was fashionable to do so.

Cousins Benoît and Paul Fouassier

In line with their natural focus in the vineyards, all fermentations take place spontaneously, with no additions of yeasts, enzymes or other fermentation aids. The cellar is equipped with temperature-controlled stainless steel, wood, concrete eggs and amphorae of differing shapes to play with. Paul explains that the single vineyard wines are fermented in stainless steel for a pure expression of the site, whereas reds are put into large old oak or amphorae. All white wines age for at least 10 months and will usually undergo malolactic fermentation, as no sulphur is used during fermentation or ageing. The firm, flint-based 'Les Romains' cuvée ages in concrete, which Paul describes as providing a good mix between stainless steel (which is impermeable to air) and amphorae, which would give it too much oxygen. 'Using concrete stops the wine being too closed in its youth,' he says. Amphorae are preferred for red wines, which age for up to 18 months before release. All of the red wine grapes are handpicked, as are a portion of the whites. The decision to machine harvest the remainder is based on the practical difficulty of hiring temporary workers, a common complaint in Sancerre and indeed across much of the Loire Valley.

Quality is high and consistent at Domaine Fouassier and it's clear the jocular cousins enjoy both their work and working together. The wines show expressive, mineral fruit. White cuvées – especially those grown on flint – show a noticeable and attractive twist of bitterness on the finish. Paul says, 'This is a mark of what we are trying to do.'

Domaine Henri Bourgeois

Chavignol

www.famillebourgeois-sancerre.com

It's hard to miss the Bourgeois family in Chavignol. There's a large and well-run tasting room (open daily), a hotel and a restaurant, all in a village so small you'd hardly expect to find any of these things. The family has had plenty of time to make its mark, mind you. They've been here since 1696, but it wasn't until the 1950s that wine became the focus of the business, under the leadership of Henri Bourgeois. His son Jean-Marie, now in his eighties, tells me that at this time there were a mere 2.5 hectares of vines, the remainder of the agricultural activity being directed towards cereals to feed rabbits, chickens, goats and cows. Little by little, viticulture grew in importance and the village became the most reputed in the area. 'And then we had to get rid of the goats,' he says, 'because you couldn't have tourists come to visit with all that mess and smell everywhere.'

Sancerre pioneer Jean-Marie Bourgeois with his eightieth birthday wine

Today it is Jean-Marie's sons and nephew who run the business and there is plenty for them to do. Jean-Marie's son Arnaud explains that they manage 134 blocks spread across 10 of the 14 Sancerre villages. They own around a third of these and have long-term contracts for the remainder. They also produce wines from Pouilly Fumé, Menetou-Salon, Quincy and Châteaumeillant, and have an estate in Marlborough, New Zealand (Clos Henri). Most of the Sancerre wines are certified organic.

There are some notably high-quality single-vineyard Sancerres from this estate. Arnaud explains the keys to the house style. Naturally, picking date is of utmost importance: 'Too early and you get excessive pyrazines [green, herbal aromas]; too late and you are in the realm of terpenes [heavier floral and citrussy notes],' he says. And the pressing has to be slow and gentle, 'to get the best acidity.' There is never any skin contact, he says, because 'we don't want anything to get in the way of the minerality.'

Les Côtes aux Valets comes from a single hectare in the south of the appellation. It is fragrant with notes of green citrus and peach, zingy acidity and appetizing ripeness. The wine ferments in stainless steel before ageing on lees in large oak barrels for 8 to 12 months. Le Cotelin is a single vineyard planted on Kimmeridgian marls. Partial barrel fermentation gives this wine a hint of tannins to complement the chalky texture, honeysuckle notes and mineral finish. Jadis is a wine made, says Arnaud, 'in the style of our grandparents' from the Kimmeridgian marls of Chavignol. Unfiltered, it offers a harmonious, fruity palate with mineral body and substance.

La Bourgeoise is a wine from grapes grown on flint. Arnaud describes the wine as 'like a really interesting person who doesn't say much.' He adds, 'Very few people bottle pure flint wines, because they don't open up for a few years.' Flint comprises only a small proportion of the land in Sancerre, but it often gives wines of arresting interest, such as this one. Fermented partly in barrel, it has a smoky character with very assertive, but nevertheless ripe, acidity. It's a wine built entirely around this acid structure. While it needs time, it's definitely worth waiting for, and this is a wine that can last and last. A 1985 tasted recently with Jean-Marie showed as few signs of tiring with age as the man himself. It had notes of honeysuckle, truffle, smoke and toast. Another wonderful flint-grown wine is Les Ruchons, which comes from vines planted in 1970. It is mostly fermented in barrel and features a highly mineral palate with appealing bitterness and a long, juicy finish. From the most famous vineyard of Sancerre, La Côte des Monts Damnés is another wine that has been shown to stay the course – I have tasted perfectly delicious 30-year-old examples of this wine – although this is an early ripening site that makes wines to enjoy from a young age. The warm, south-facing Kimmeridgian marls of this steep vineyard have been farmed by the locals for a thousand years.

The domaine's reds are worthy of note, especially Le Graveron, which grows on a 1.3-hectare plot at the top of the Monts Damnés vineyard. Hand-made and Burgundian in style, it has perfumed, savoury fruit and supple, silky tannins.

Everything is done well at this estate, from the wines to the welcome in the vineyards, tasting room, hotel and restaurant. It should be on every visitor's list when in the region.

A perfect match

The classic match for Sancerre Sauvignons is the local cylindrical or disc-shaped goat's cheese, *crottin de chèvre*, especially from the village of Chavignol, where goats used to outnumber vines. These little cheeses can be soft, fresh and slightly creamy, or aged to a harder texture when they show a distinctive 'goaty' tang and a mildly nutty character. But at any age (and any time of day, it seems), Sauvignon from Sancerre hits the spot with this local speciality.

Goat's cheese is the perfect match for Sancerre

Domaine Paul Prieur

Verdigny

www.paulprieur.com

Luc Prieur is the eleventh generation of his family to farm the family's vines in the heart of the Sancerre appellation, taking over the estate in 2012 when he was just 22 years old. 'The production has remained artisanal,' he says, 'we work by hand.' The domaine is 19 hectares, including

5 hectares of Pinot Noir. When Luc arrived, there was only one cuvée each made for red and white. Today there are four whites, three reds, and a solera rosé, plus a sparkling Vin de France.

Luc's aim is, he says, to produce wines that are 'as natural as possible, but in a scientific way.' To do this, he harvests everything by hand, keeps a 'very clean cellar' and doesn't add anything to the grapes before fermentation. There is no filtration, but he does add sulphur dioxide at bottling. 'I don't like "natural" wines at all,' he says.

The domaine Sancerre Blanc is precise and clean, crisp and pure, a pristine expression of the appellation. It comes from a blend of different sites that represent all the soils of Sancerre. Fermented and aged in stainless steel, it is by far the largest cuvée of the domaine. The other wines spend time ageing in large oak. The Monts Damnés wine comes from the south-east side of the famous vineyard, which ripens a little later than the majority with south to south-western exposure. Nevertheless, it has the open ripe fruit of the site combined with bright acidity from the Kimmeridgian marls. Pieuchaud is a wine that is grown on flint. In Luc's view, this soil gives 'a less complex but more direct, straight expression' of Sauvignon Blanc. He works hard to get the maturity exactly as he wants it for this site, compared to the Monts Damnés: 'On Monts Damnés you can do what you want and it will always be good, but I don't want to make a caricature of *silex*.' It is a mineral wine, with firm acid tension.

The red wines age partly in wood and partly in amphorae. The main cuvée of the domaine, Grains de Pinot, is partly whole-bunch fermented to give an aromatic, textured wine. 'We have more limestone here compared with Burgundy,' says Luc, who studied there, 'We have more acidity, so we can look for higher levels of maturity.' The wine is very ripe, but alive and fresh. In contrast, the Les Pichons wine is entirely destemmed. This is a glossy and seductive, black-fruited wine that comes from low-yielding vines on a high site next to the Monts Damnés vineyard. It retains the characteristic balance and freshness of the domaine.

Domaine S and J P Agisson

Sury-en-Vaux

Jean-Philippe Agisson's domaine is just a few years old, but he has already proved his expertise as a winemaker locally: he worked for 12 years alongside Alphonse Mellot before moving across the river to become cellarmaster and winemaker at Domaine Didier Dagueneau, where he has worked since 2018.

The confident and capable Jean-Philippe Agisson

His own small venture is in the quiet village of Sury-en-Vaux, and consists of 1 hectare of vines in Sancerre, where he makes a red and a white wine, and half a hectare in Pouilly Fumé. The grapes are hand-picked and taken in small cases direct to the press. There is no destemming here, because Jean-Philippe believes it risks damaging the skins. Every decision here is considered: 'Wine is a sum of the details,' he says. When ageing his wines, he prefers to use smaller barrels for the Pouilly Fumé to give it greater richness, and larger ones for the Sancerre, to make them more upright. The wines all ferment and age in oak.

His first vintage for Sancerre was 2017. These are firm, mineral wines that can benefit from a few years in bottle to reveal themselves. A four-year-old wine was, he said, 'starting to show the terroir.' Nevertheless, the young vintages of the wine – named Equilibre – show expressive fruit and are wholly enjoyable. For the red wine, also called Equilibre, Jean-Philippe uses a percentage of whole bunches (which changes according to the vintage), but is not trying to emulate the Pinot Noirs from further east. He says, 'Lots of vignerons want to make burgundy, and use lots of small oak. But this is not the identity of Sancerre. It should be less extracted, fine and elegant.' Jean-Philippe's light touch includes using a watering can to drench the cap instead of anything more extractive, such as *pigeage*, or punching down. This is a very fruit-forward wine that displays floral and black cherry fruit, a touch of spice and delicate structure. It's deliciously understated.

Since 2020, Jean-Philippe has also made a Pouilly Fumé, La Belle Endormie. The label for this wine, which translates as The Sleeping Beauty, features the back of a woman waking from her slumber and looking into a mirror reflecting back an image of the hill of Sancerre, an image which, says Jean-Philippe, represents the 'sleeping beauty' status of Pouilly Fumé today, where land remains far cheaper than Sancerre (he paid 30 per cent less for his land there than in Sancerre).

These are precise and delicious, very well-made wines that are available only in small quantities. Jean-Philippe's reassuring attention to detail and his experience of working with the very best are reflected in these wines.

An essential tool for Jean-Philippe Agisson's delicate Pinot Noir

Domaine Tabordet

Verdigny-en-Sancerre
www.domaine-tabordet.fr

The Tabordets are a family of farmers with expertise in rearing cows and goats as well as cereals and vines. Marius Tabordet, now managing the wine side of the business, says he'd like to have some of the goats, but recognizes 'you can't do everything'. His hands are pretty full as things stand. The domaine's vineyards consist of 16 hectares of Pouilly Fumé vines and 5 hectares of Sancerre (3 white, 2 red), and he makes 12 wines, 6 in each appellation.

Marius's father, Pascal, was one of seven children, each of whom was given one part of the business to manage. Pascal got the vines, which have now been passed on to Marius, although his father still helps with the commercial side of the business.

The domaine was set up by Marius's grandfather, Raymond, in the 1960s, initially as a 4.5-hectare Sancerre-only enterprise. In 1981 the business expanded into Pouilly Fumé. Marius trained in New Zealand with John Forrest, and in France with François Villard in Côte Rôtie and Alphonse Mellot in Sancerre before returning to the domaine in 2008. At this time, it was run wholly conventionally. He gradually persuaded his father to start cutting out herbicides and from 2010 began moving slowly towards organic viticulture. The domaine was certified organic in 2018, and his father was against going any further, but a determined Marius went ahead with his plans to convert the entire estate to biodynamics and is happy to report that along the way he also converted his father to his way of thinking.

The Sancerre Domaine white is a zingy, ripe and pure wine, fermented in stainless steel, with no sulphur dioxide added until bottling. This, like all the whites, has undergone malolactic fermentation, but it is not readily detectable. 'I like the creamy, pastry side of malolactic,' says Marius, 'but I don't want to smell it.' The Pouilly Fumé Domaine is made in a similar way. It is very expressive, something that comes easily in Pouilly, according to Marius.

The cellar contains rows of spotless stainless steel tanks, 'installed by my father,' says Marius, 'now I wouldn't choose to do this.' He is more of an experimentalist, especially when it comes to the effects of fermenting and ageing in amphorae and oak. One wine, Aventure, is matured for three years in 600-litre oak barrels, then bottled with no filtration or sulphur dioxide. 'Year one, it doesn't move,' Marius says, 'year two it is oxidized, all over the place – and then in year three, it only gets better.' But while he isn't afraid to take risks, he expresses a fear that must be familiar to many vignerons: 'The moment of harvest is the hardest,' he says, 'you find yourself in the middle of the vines and all alone.'

Marius has been described by others as the most talented young winemaker of his generation and, even if he is experimenting in some of his wines, his underlying sensitivity means they all seem to turn out remarkably well. Certainly, there is nothing boring here, much to be excited about and a very bright future ahead for Domaine Tabordet.

Domaine Vacheron

Sancerre

Based in the centre of Sancerre, Domaine Vacheron is run by cousins Jean-Laurent and Jean-Dominique Vacheron. The family has been in Sancerre since the early twentieth century but, like most families, practised a polyculture that included goats and cows as well as vines. Now focused solely on wine production, the cousins farm 75 hectares of land across 15 different vineyard sites, producing a range of excellent white and red wines.

The family's vineyards are all within close reach of the village, on either side of the fault line that divides soil type and geological time. Standing at the lowest point of the valley that marks the line of the fault just south of Sancerre, Jean-Laurent explains: 'To the left [west], the soils are 140 million years old; to the right, 40 million years.' The difference is clear to even the most untutored eye. The young flint soils look, feel and even smell different from the older clay-limestone ones just two steps away. Touring the vines with Jean-Laurent, it is clear that he knows the land intimately – including who owns every vine: 'Here is Fouassier,' he gestures, 'and this is Alphonse Mellot. Here, our vines are just next to Joseph Mellot.' The peculiarities of each site – slope, soil and aspect – will determine the character of the wines they make. The precision with which he describes the composition of the soils makes viticulture feel like an exact science.

As well as a blended house cuvée that is a good representation of the classic Sancerre style, the Vacherons make five different *parcellaire* Sancerres, each with its own unique character. The grapes in 'Le Paradis' grow on a steep, south-facing site of hard, Oxfordian limestone,

Evening sun lights up the vineyards in Sancerre

with many small stones (*caillottes*). Jean-Laurent points out the 'bonsai vines' in this dry vineyard, that produce floral, mineral wines. Nearby 'Chambrates' has similar soils and exposition, but on flatter land with a greater proportion of clay, which gives a fleshier style. 'Guigne-Chèvres' – also very close by – is a north-east-facing plot on clay-limestone that gives always-elegant wines with texture, fruit and power. The grapes in 'Les Romains' grow on a steep, south-west-facing slope of flint, and yield age-worthy wines deeply evocative of their origins, with dusty mineral notes and chiselled structure. In the steep and warm, easterly vineyard on the other side of the town, the Pavé vineyard has clay-based soils. Here, grapes develop thicker, more golden skins and the wines they yield are powerful, firm and spicy.

An additional tiny cuvée of some 300 bottles a year is made from ungrafted vines planted on a sheltered, south-facing slope within the walls of Sancerre itself (and therefore outside the boundaries of the appellation). This Vin de France is called 'L'Enclos' and it is planted on flint. 'The idea behind this is to see what the characteristics of Sauvignon are in the wild,' says Jean-Laurent, postulating that this gives wines that have more of the thiol (citrus) and terpene (floral) characters of Sauvignon Blanc and less of the green pyrazine characters that can come from rootstocks and which Sancerre producers are so keen to avoid. Other ways of tackling the dreaded pyrazines elsewhere include tactical deleafing, depending on the sun exposure of each plot. For new plantings the Vacherons produce a massal selection of vines from their own plots, and work with a specialist nursery in Alsace to provide their own rootstocks as well.

Red wines make up a small proportion of production but are among the very best of the appellation. The flagship cuvée, 'Belle Dame' is a complex, scented and satisfying wine of immediate appeal, grown on flint soils. 'Les Marnes' is gorgeously silky and elegant, made from old vine fruit grown on Kimmeridgian marls. The entry-level Sancerre Rouge is always a beautifully fresh and expressive wine, gently smooth and polished. The cousins completed an ambitious renovation of the cellars in 2021 – especially ambitious when you consider the logistics of working with heavy machinery in the narrow, winding streets of Sancerre – and have now installed nine large, conical wooden vats for their red wine fermentations and a beautiful tasting room that overlooks the roofs and turrets of the town and the vineyards beyond.

Domaine Vincent Gaudry

Sury-en-Vaux

vincent-gaudry.com

There's nothing ordinary about Vincent Gaudry. Born and raised in Sury-en-Vaux, where he still lives today, Vincent is a winemaker who lives and breathes his terroir to such a degree it's hard to imagine he is anything but an integral part of it. He describes his philosophy as 'very traditional'. His vineyards are horse-ploughed and the grapes harvested by hand. 'We are inventing nothing here,' he says. He talks a lot about *les anciens* – the old vignerons who have taught him indefinable things, such as 'seeing the presence of the invisible'. He also talks a lot about energy, so it's no surprise to learn he was the first vigneron in Sancerre to be certified biodynamic (in 2004), a philosophy he claims was heavily criticized at the time and that even today, very few people understand.

Vincent farms vineyards that have been in his family for generations, although he is the first to make a living purely from wine. Things did not get off to an auspicious start: his first vintage, 1991, was a disaster. 'Zero grapes – worse than 2021,' he says. Things could only get better, and indeed they have. The last time I saw the dungaree-clad vigneron he was being feted in a local restaurant by a gaggle of well-dressed New Yorkers, all clamouring for a piece of the Gaudry mystique.

The wines are undeniably delicious. Vincent claims he is not looking for complexity, but subtlety. He manages to achieve both. The entry-level Le Tournebride is a fine-textured and elegant expression of the three terroirs of Sancerre. It comes from the younger vines of the estate and is fermented in stainless steel, enamel-lined vats and concrete. Mélodie de Vieilles Vignes is a wine from vines of 60 years or more that grow on clay-limestone soils. It has a mineral and floral intensity with a rounded mouthfeel and a long, beautifully textured finish. There is an attractive herbal edge to the fruit, and Vincent comments that here he can taste the sap of the vine. The Constellation du Scorpion wine might be considered to be his own cuvée, since this is his birth sign. Grown on flint soils, it has the characteristic lifted smokiness and sharp salinity of wines grown on flint. Light and elegant on the palate, it finishes with a distinct and savoury note of white pepper. From the same vineyard, À Mi-Chemin is a wine from grapes grown at the top of this south-west-facing slope. Fabulously aromatic and wonderfully balanced, it has juicy, lightly herbal fruit and a powerful impression of the warm stones of this site – a true *vin de terroir* with intensity and mineral finesse.

Another wine, Pour Vous, comes from 50-year-old vines that grow on clay-limestone soils. This is aged for two years in old 228-litre oak barriques: 'The old texts say Sancerre used to be aged like this,' Vincent says. There is no evidence of oak, only juicy, star-bright fruit. The vineyard always produces lovely fruit, says Vincent, so good you always want to share it – hence the name 'For You'. Describing this wine, Vincent says, 'It has a lot of force and energy. You need to switch off your brain and listen to your body – and when you swallow it, it's like the sun radiating out from your being. Every place has its energy and this is what we are trying to capture in the bottle.'

Domaine Vincent Pinard

Bué

www.domaine-pinard.com

This longstanding family domaine is currently in the hands of Clément and Florent Pinard. Their father Vincent handed over the reins in 2000, but remains a familiar presence at the winery. With 17 hectares spread over 40 parcels, the brothers have their work cut out to manage it all and say they spend 90 per cent of their time working in the vines. The vineyard is certified organic and farmed biodynamically.

They produce 13 wines: eight whites, four reds and a rosé. Each provides a different expression of the local soils. It is a fascinating exercise to taste through these wines, although to be honest the differences do begin to blur after the forty-eighth wine of the afternoon (no joke). But as much as the nuanced terroir differences, it is perhaps the similarities that distinguish the superior quality of these wines. The work is meticulous and considered: 'Nothing is systematic,' says Florent, yet there is systematic attention to every aspect of the grape growing and winemaking here, to make the most of what nature gives them. Much of the work is manual, and many of the vines are more than 50 years old. The grapes are all hand-picked and undergo gentle handling in the cellars. There is no fining or filtration before bottling, and a very modest use of organic sulphur in the winery. The brothers are experimenting with lower levels of sulphur or even none at all, as with a recently created cuvée, Narpi.

The reds, a minority of production, are of notable quality and a reference in Sancerre. Charlouise, named after a contraction of the names of the brothers' grandparents and produced from the oldest vines, is consistently outstanding, but it's hard to find a bad wine anywhere here. The whites are also impeccable and again, it is the older vine cuvées, such as Grand Chemarin, Petit Chemarin, Le Château and Chêne Marchand that stand out, showing gorgeous depth of flavour, texture and terroir expression. Spend time with the Pinard family and one senses a constant quest for the next small notch of quality improvement, with a clear vision of where they are heading. 'In wine,' says Clément, 'it is important to have conviction but not to be inflexible.'

François Cotat

Chavignol

When wines are obtainable only on allocation (although – don't tell anyone – they have also been spotted in the local cheese shop), it's clearly a sign of success. But François Cotat wears his renown lightly. He is also extremely camera-shy ('No you can't take a picture – just look at me!'). With no pretence, and certainly no ceremony, we duck straight into his tiny cellar tasting room, under the house on a narrow Chavignol side street. It could not be more low-key. Lit by a bare bulb, the room contains a simple table and chairs, a jumble of old photographs, newspaper clippings and posters, walls lined with racks of wine and a few bottles uncorked and ready to taste. They may have been opened an hour, a day or a week earlier. It hardly seems to matter. François knows his wines, even if he appears to be as in the dark as anyone as to why they have such legendary capacity to age.

The grapes come from a small selection of some of the finest plots in the Sancerre appellation although, as he points out, this is Chavignol wine, and one gets the impression that if he had his way there would be no mention of Sancerre anywhere. Look closely at a bottle of François Cotat wine and 'Chavignol' is the larger of the two words – these things matter – and the only one printed on the cork.

He both owns and rents vines ('Nothing ever comes up for sale here') on the slopes surrounding the village. The domaine is only 5 hectares in total. These include the prime *lieux-dits* Cul de Beaujeu and Monts Damnés. These steep, south-facing slopes are necessarily worked entirely by hand: brutal, physical work. He also farms a hectare on 'La Grande Côte' (also known as 'La Côte d'Amigny'), just south of Chavignol, where he grows both Pinot Noir and Sauvignon Blanc.

Viticulture is not organic, but, he says, is generally non-interventionist. In the cellar, everything is natural. Fermentations take place with indigenous yeasts, and there is no fining or filtration. François says he follows the moon to decide the right moment for bottling. Contrary to other top producers, there is no long lees ageing here, three months at

The next best thing to a photo of François Cotat: his tasting room

most. It all sounds very efficient: 'Eight months from harvest to bottling,' he says, adding, 'I don't like long *élevage*; it tires the wines.'

Key to the Cotat style is ripeness allied with mineral intensity. Picking dates are often late, although such decisions are based, he says, 'entirely on taste'. Old vine fruit picked at optimal ripeness gives wines of depth and concentration, but there is no excess discernible here, merely a precise, balanced expression of great vineyard sites.

There are four white cuvées: Caillottes, which comes from younger vines, Grande Côte, Cul de Beaujeu and Monts Damnés. He also makes a small quantity of deliciously soft and delicate but highly complex rosé that comes from 60-year-old vines, and an even smaller quantity of red, 'Chavignol'. Sometimes the wines are denied the Sancerre appellation because they contain a few grams of residual sugar or, in the case of his red wine, because he might block the malolactic fermentation required to claim the Sancerre appellation, 'so that the wine lasts longer.' He is clearly wholly unconcerned by the actions of the powers that be, and it gives him the opportunity to call his red Vin de France Chavignol. It probably goes without saying, but his customers appear not to care either.

Gérard Boulay

Chavignol

The Boulays can trace their Chavignol roots back as far as 1379 so perhaps it is no surprise that they have some of the best terroirs in the village – nor that Thibaut Boulay, who runs the domaine with his father, Gérard, chose to specialize in studying history as a career, something he manages to do alongside running the domaine. He has also written a definitive book, *Les Terroirs Sancerrois* (only available in French), providing a detailed analysis of the vineyards, and offering a selection of walking itineraries around Sancerre for those wishing to get up close and personal with the vines. Thibaut knows this land like the back of his hand; walking around the Monts Damnés vineyard with him is a masterclass in natural history. Every few metres he stops to point out a different feature of the landscape – a rare plant or insect, a fossilized seashell, a viewpoint. Chavignol is the heart of wine growing in Sancerre and has always produced white wine, says Thibaut.

The family's parcels are all on the Kimmeridgian marls of Chavignol and include 2.5 hectares of the Cul de Beaujeu and 1.2 hectares of the Monts Damnés vineyards. Thibaut's father, Gérard, has been at the helm since 1972. The biggest challenge, says Gérard, is that everything on the slopes needs to be done by hand. Most of their time is spent in managing the vineyard and very little in the cellar.

Gérard prefers a riper style of wine: 'I like good ripeness and less acidity,' he says. The fact that he has some of the ripest sites and 'lots of slopes' encourages this style. The Monts Damnés faces due south; the wine from here is rounder and lower in acidity than the Cul de Beaujeu, which faces south-east. Gérard prefers the Beaujeu, however, because of its greater ageing capacity. To prove the point, he pours a 2009 Clos de Beaujeu. With aromas heading towards diesel notes, this is a fascinating wine with lovely balance and length, peachy, pineapple fruit and juicy acidity. It is ageing beautifully at 15 years old.

The Boulay wines are made using grapes from vines that are on average 50 years old. The winemaking features indigenous yeasts and all wines except the Tradition cuvée ferment in wood. The Comtesse Sancerre cuvée comes from the best, lower, part of the Monts Damnés vineyard, on pure Kimmeridgian marls (the upper part has harder, Portlandian limestone). This is a ripe and open-knit style, with lovely fruit concentration. It can also age beautifully, a 2012 Comtesse showing great complexity and mineral length. This wine was nowhere near its peak, either, and also nothing like a traditional expression of Sauvignon Blanc grapes. A small amount of red is made for the cuvée Oriane, a ripe and smooth Pinot Noir with savoury, serious fruit and gentle tannins.

With so many centuries of history behind them, it's fair to assume the Boulays will be making wine here for generations to come.

Jean-Max Roger

Bué

www.jean-max-roger.fr

You can't walk far down rue St Vincent, the main street of Bué, without tripping over a winery. It's a fair bet that all 300 inhabitants of this little village come from families that have been steeped in wine for generations. The Roger family has been here since the seventeenth century, and possibly even earlier. Today, the winery is run by ebullient brothers Thibault and Etienne Roger. Their father, Jean-Max, took over the few hectares of family vineyards in 1972 and steadily increased the scale of the business over the next 30 years. The brothers now farm 35 hectares in vineyards across Sancerre, Menetou-Salon and Pouilly Fumé.

Thibault and Etienne, the brothers Roger

Entering the small house off the main street one finds a simple tasting room with a table and bench, terracotta-tiled floor and oak-beamed ceiling. From here a side door opens to reveal an unexpectedly modern and sizeable winery, where lines of stainless steel tanks fill every inch of space. These are used to ferment the larger cuvées of white wines. A more traditional fifteenth-century cellar houses 400-litre barrels which are used for their Vieilles Vignes (from vines aged 40 years or more) and prestige cuvées. After fermentation, the wines typically spend a few months on fine lees for added texture.

The various Sancerre cuvées reflect the soils, individual plots or *lieux-dits*, and age of the vines. Thibault explains: 'The *caillottes* give more expressive aromas, while the *griottes* soils are finer and more elegant. *Terres blanches* soils give more structured wines.'

The Roger brothers offer a fine range of wines that reflect their terroirs. Menetou-Salon bottlings are frank, early-drinking Sauvignons with clear varietal typicity, while the Sancerres show mineral, textured fruit and clear acid structure. The *lieux-dits* of Grand Chemarin and La Côte de Bué are particular favourites. These wines, from vineyards in the village, enjoy the extra care of hand-picking and are partially fermented in barrel. The brothers also make a small quantity of elegant red Sancerre Vieilles Vignes.

Lucien Crochet

Bué

www.lucien-crochet.fr

This family estate of 36 hectares is planted 26 hectares to Sauvignon and the remainder to Pinot Noir, mostly in and around the village of Bué. The grapes grow on the 'Terres Blanches' clay-based limestones, and the stony soils of *caillottes* and *griottes*. Winemaker Gilles Crochet (descended from two Luciens) managed the domaine with his wife Laurence until his death in February 2023. Gilles's goal was to show the terroir, rather than the aromatic side of Sauvignon Blanc which, as he said, 'anyone can do.'

The winemaking has now been taken over by Clément Bizet, a 26-year-old from Bué who had been working alongside Gilles. Compared to other wineries of a similar size, the domaine produces relatively few cuvées, just three whites and two reds. The whites come from a blend of sites of different exposures and altitudes, but all are from the stony limestone soils of *caillottes* and *griottes*. The reds are grown partly on these soils, but also on clay-based Kimmeridgian marls, which give the wines more body and structure.

Winemaking begins with hand-harvested grapes for all the wines, which then undergo fermentation with native yeasts in temperature-controlled vats. Whites benefit from long lees ageing, first on gross lees and then on fine lees, for around 17 to 19 months. Some, such as the Calcaires and Chêne Marchand wines, are aged in 600-litre oak barrels and blended in, to give roundness and softness, while the Cuvée Prestige sees a little more new oak. The Chêne Marchand is particularly appealing, with gorgeous chalky texture. La Croix du Roy red is a delicious wine, with clear fruit and well-judged oak. The destemmed grapes spend 30 days on skins before being transferred to 228-litre Burgundy barrels and 600-litre *demi-muids* for a year of ageing in wood, then time in tank before bottling. Hopefully, Gilles's successor will continue the fine work of his predecessor to build on his legacy of excellence at this domaine.

Other producers to try

Domaine Paul Thomas, François Crochet, Domaine Delaporte (with a tasting room in Chavignol), Pascal Reverdy, Château la Rabotine.

When in Sancerre

Places to taste wine

The Maison des Sancerre (www.sancerreaop.com/la-maison-des-sancerre) is a mine of information on everything to do with Sancerre wine. It offers visitors an excellent 4D presentation of the history and practices of the vineyard and winemaking up to the present day, with some surprises along the way. There's also a shop with posters, books and wine paraphernalia and a pleasant terrace overlooking the town, from which a variety of wines can be sampled. On Friday evenings in summer, ('*Vindredis*'), local producers set up tasting stations on the terrace for sampling.

A few individual producers sell their wines from retail outlets in the middle of Sancerre, including Alphonse Mellot and Domaine Berthier, and the Berthier brothers also own the centrally located wine bar, La Banque (www.labanque-sancerre.fr), which is worth a visit. In 2022, four growers came together to create Vignerons La Cave, an independent wine shop selling their own wines together with a personal selection from the rest of France (www.vigneronslacave.com).

Places to stay and eat

The Panoramic Hotel is a decent hotel option in Sancerre itself, although there is a good selection of private accommodations available. Don't miss the sunset from the terrace of the Panoramic Hotel bar, which also features a very good selection of local wines (www.panoramic-hotel.com). In the pretty village of Chavignol, in the shadow of the famous vineyard of the same name is La Côte des Monts Damnés (www.hotel-restaurant-chavignol.fr), a hotel and restaurant owned by the Bourgeois family. After visiting the domaine's welcoming tasting room, it's a short stroll down the road to enjoy a tasty bite on the terrace by the village fountain. Unsurprisingly, Chavignol's most famous other export, the crottin de Chavignol, features prominently on the menu. To take home cheese souvenirs, the largest crottin producer, Dubois Boulay, has a shop in Sancerre and a larger outlet on the outskirts of Chavignol village (dubois-boulay.fr). For lunch in the vineyards, Au Bon Laboureur in Menetou-Râtel is a top choice for market-fresh produce, simply but skilfully put together. Don't be surprised if you bump into local vignerons in this popular local, which also features an excellent wine list with more than 650 references (www.aubonlaboureur.fr). Sancerre and its surroundings also have a good selection of local restaurants.

The cobbled streets of Sancerre by night

Getting around

Distances are small here, so getting around by car is easy (though the narrow streets of Sancerre itself are not for the faint of heart). To really get a feel for the vineyards though – and to appreciate fully just how steep the slopes are – there is nothing better than a bike for exploring the area. Several places rent bikes, including Cycles Sancerrois (Rte de Chavignol, 18300 Saint-Satur; 02 48 54 38 59). Electric bikes are a popular choice for tackling the hilly terrain.

Pouilly Fumé/Blanc Fumé de Pouilly – key facts

- **AOC granted**: 1937
- **Hectares in production**: 1,375
- **Annual production volume**: 79,000 hectolitres/10.5 million bottles
- **Styles**: 100 per cent white
- **Grapes**: Sauvignon Blanc
- **Soils**: Kimmeridgian marls, limestone, flinty clays

Pouilly-sur-Loire – key facts

- **AOC granted**: 1937
- **Hectares in production**: 25
- **Annual production volume**: 1,300 hectolitres/180,000 bottles
- **Styles**: 100 per cent white
- **Grapes**: Chasselas
- **Soils**: Kimmeridgian marls, limestone, flinty clays

Notable Pouilly Fumé producers

Alexandre Bain

Tracy-sur-Loire

The first time I saw Alexandre Bain was through the arms of a crowd of natural wine enthusiasts, their empty glasses outstretched to get a taste of his wines. This was at the chaotic crush of La Dive Bouteille, the annual natural wine fest in Saumur. 'You should come to the domaine,' he shouted above the din, 'I have many things for you to taste.' The calm of his *chai*, a purpose-built, functional and brightly lit space lined with an assortment of barrels and tanks, was an altogether more conducive environment to contemplate these highly unusual wines.

Alexandre is a winemaker who has a lot to say and a great passion for what he does. 'I have a deep love of wine and everything that surrounds it,' he says with feeling that perhaps comes from having had the freedom of choosing this life rather than being born into it. He grew up with a love of nature and horses, but had no background in wine. 'I just knew I always wanted to respect the earth,' he says. Discovering wine at the age of 18, he went on to study viticulture and oenology. Alexandre says it was always his intention to make natural wine, even though when he started here in 2007 everything was farmed using chemicals and harvested by machine. He began to change all this, making wines as naturally as possible and ageing them for long periods in his cellar before release.

It didn't take long for his talent to be recognized: soon his wines were on all the best restaurant lists. But in 2014 the local authorities stripped him of the right to use the Pouilly Fumé appellation on his labels, saying his wines were atypical of the appellation. Furious and indignant, Alexandre took them to court, claiming there was no established 'typical' Pouilly Fumé style and he was making wines in keeping with all the rules of the appellation. After a two-year fight, the court agreed with him and he was given the right to the appellation again. But having won his battle, Alexandre decided to snub the establishment altogether. His wines – which continue to be some of the most successful in the region – have been labelled Vin de France ever since.

Alexandre Bain in his happy place

So what makes these wines so special – and so different? Because they are made as naturally as possible, there are added risks to contend with. 'Managing bacteria is an everyday stress,' explains Alexandre, 'You need to have a strong character because when things are difficult, the human reflex is to add something.' He manages problems by taking his time, by controlling temperatures (cooler temperatures slow everything down) and watching, and tasting, and waiting for nature to take its course. The minimum length of time he has his wines in the cellar is 18 months, but this can be much longer. 'Natural wines, especially whites, need *élevage* to stabilize,' he says.

The wines are quite unlike anything else and it's easy to forget that they are all made entirely from Sauvignon Blanc. The grape seems almost irrelevant, subsumed into the expression of the place and the mind of the person who has curated them into being. 'I don't like explosive wines,' says Alexandre, 'I'm looking for volume and length – a "grande bouche".' The samples I tasted from his brightly painted concrete tanks all shared extraordinary length and depth of flavour. Pierre Précieuse has flavours of an entire fruit basket, from peaches and pears through to tropical mango and papaya; in other tanks there were delicate white flowers, hawthorn blossom, honeysuckle, almonds and minerals; in L d'Ange there was tangy, salty acid, smoke and bitter complexity. 'Bitterness challenges the drinker,' says Alexandre, 'it always makes the hair on my arms stand on end.' La Levée is another wine unlike any other: so much of everything – intensely fruity and smoky, with a wild mineral acidity, and everything perfectly in balance. It's more an experience than a drink and it's easy to see why sommeliers have a field day with these wines, which almost defy description.

Alexandre's wines are unique, fascinating, spine-tingling. He says, 'It's an artisanal product. I make it with my hands, my heart. If you like it, that's great, but if you don't like it, it's still me.'

Domaine Didier Dagueneau

Saint-Andelain

Didier Dagueneau's name is the most famous – and infamous – in the Pouilly Fumé appellation. The wild child of Pouilly turned his back on his winemaking roots in his youth to become a professional racing driver, but when he did finally decide to give winemaking a go, he blazed his own spectacular trail, doing everything, uncompromisingly, his way. Starting with little more than a hectare of land, he adopted a meticulous approach to grape growing and winemaking, and the results were wines that everyone wanted to buy.

It might not sound unusual today, but in 1982 it was revolutionary to slash yields, use a horse-drawn plough in a vineyard and adopt a terroir-focused approach to wine growing. The cuvées Pur Sang and Silex that Didier Dagueneau created in the mid-1980s are still icon wines today, and the name Silex has been borrowed by countless vignerons since to highlight their own *silex* (flint) terroir. Today it is Didier's son, Louis-Benjamin who continues the legacy of his father, who died prematurely young in a light aircraft accident in 2008.

There are five different cuvées here, and Louis-Benjamin, who had already started working with his father at the time of his death, uses the same winemaking methods for all. Hand-picked grapes are destemmed and fermented in barriques, of which a quarter are new. The wines age on their lees for a year, then in stainless steel tanks for eight to ten months before bottling. Louis-Benjamin is also experimenting with using glass 'wine globes' which allow for zero oxygen ingress.

Blanc etc is a wine from a selection of different soils that include mostly Kimmeridgian marls and around a third flinty clay. It is pure, fresh and very mineral, with a long, flinty finish and light tannins. Pur Sang comes from a south-facing, three-hectare plot of flinty clay with flint pebbles. It has an initially rich palate that gives way to precise, sweeping acidity and a fine, chalky texture. It has wonderful finesse. Buisson Renard is a west-facing site with richer clays, giving wines that feel a little riper and more powerful, but that retain a zesty elegance. It is aromatic, floral and quite open when young. Silex, in contrast, can be rather closed in its youth. Firm and mineral, with acidity that runs through it like an arrow, it is from the vines closest to the winery in Saint-Andelain, a high site with poor soils of flinty clay. There is also a Sancerre from the Dagueneau domaine. This comes from the Monts Damnés, the renowned south-facing slope above the village of Chavignol.

While Louis-Benjamin has continued to produce wines in the image of his father, he is always looking for ways to improve. To combat climate change he has planted some Petit Meslier in the vineyard, an ancient variety known for its ability to retain acidity in warm conditions. He has also introduced glass wine globes for maturation of his micro-cuvée, L'Asteroïd, a wine from a 0.17-hectare plot of ungrafted vines planted in 1991 in the heart of the Pur Sang vineyard. He has recently removed the wines from the Pouilly Fumé appellation, so they are all now labelled as Vin de France.

Domaine du Bouchot

Saint-Andelain

www.domaine-du-bouchot.com

Antoine Gouffier grew up in the vineyards of Pouilly Fumé, but spent a decade overseas before putting his considerable energy and drive into making a name for himself as a vigneron back home. He has travelled widely, latterly spending 6 years in South Africa. Returning to his roots in 2018, he bought Domaine du Bouchot, working one vintage with the former owner, and then taking over on his own in 2019. His first solo vintage was 2020.

Part of the appeal of buying this estate, Antoine tells me, is that it was the first farm in Pouilly Fumé to be certified organic, back in 2009, and the vines were already in conversion to biodynamics. Working on a small scale, he now farms 10 hectares of the estate's 37 and makes 3 cuvées of Pouilly Fumé from the different terroirs that surround the winery. Everything is harvested by hand, although he grumbles that it is always difficult to find people who want to do vineyard work.

Terres Blanches is a wine from grapes grown on Kimmeridgian marls. It is a ripe and floral wine, mineral and tangy. Most of the grapes ferment in stainless steel and just a small percentage in oak. It is a multilayered wine that combines mealy richness and zesty freshness. Caillottes, which comes from rockier limestone soils, has a lot of substance, but also more tension. 'This terroir always gives a lot of acidity,' says Antoine, 'whereas the Kimmeridgian marls give more fruit.' It's a beautiful wine, with a splash of creamy oak. 'What I'm happy about is that it doesn't taste of Sauvignon Blanc,' he says, with no hint of irony. MCMLV comes from old vines of 60 years or more, growing on Kimmeridgian marls. Half of the grapes are macerated for a few days on their skins, the rest pressed directly, and half age in 500-litre barrels on lees for a year. Unfiltered, this has full, toasty richness, lovely texture and mouthwatering minerality. It finishes with a floral, saline touch.

Antoine also makes a Pouilly-sur-Loire Chasselas – Mon Village – which outperforms the appellation. There is skin contact at the beginning of the winemaking here and lees ageing at the end, all in old barrels. It's an unfiltered, creamy delight. He also makes an orange wine – easy to identify, as it's called Orange – for which the juice is in contact with the skins for six months in barrel, where it ferments while Antoine periodically rolls the barrel. The finished wine is transferred for further ageing in concrete eggs. It is a fruity, citrussy, gently textured wine with a mineral finish.

Antoine Gouffier has quickly become a name to watch in Pouilly Fumé, crafting harmoniously delicious wines with flair.

Domaine Jonathan Didier Pabiot

Pouilly-sur-Loire
www.jonathanpabiot.fr

Jonathan Pabiot is a fifth-generation winemaker who took over his father's 20-hectare domaine in 2010, quickly changing the direction of the business. 'We didn't keep a single customer,' his wife, Nina, explains, 'because the prices went up and we stopped making Sauvignon Blanc-style Sauvignon Blanc.'

The winery is now dedicated to making terroir-driven wines from low-yielding vines. A third of the land is around the winery, close to – and within sight of – the Loire, something which gives a very humid climate in the vines. According to Nina, this microclimate is also a source of minerality in the wines, because the grapes experience a lot of heat in the afternoon, followed by cool freshness in the morning.

The largest cuvée of the domaine is Léon, with a terroir of clay-limestone and flint. Part of the harvest ferments in concrete, part in stainless steel, and part (just 20 per cent) in oak. It's a stylish wine, intense and mineral, with a smoky touch and a hint of creamy toastiness. Another wine, Aubaine, grows on pure limestone soils. This is all elegance, with citrussy fruit, a rich, floral, leesy palate and a hint of bitterness on the finish. The wine is fermented in concrete tanks and left on its lees for 18 months. 'We leave it alone,' says Nina. The Luminance wine is made in the same way as Aubaine, and comes from Kimmeridgian marl soils. It has an uncompromising firmness with great depth of wholly savoury flavour. It's a connoisseur's wine that will age beautifully. The wine has concentration of flavour from the low yields and excellent terroir. The Eurythmie wine offers a change of pace: this is a flint-grown barrel-fermented and barrel-aged wine that feels altogether bigger and more generous, with textured oak and luscious fruit. More open in style, it is no less complex but could be approached earlier. The range also includes a Pouilly-sur-Loire Chasselas with an attractive wet-stone character, and a natural wine, Utopia. The grapes for this come from the same parcel as for Luminance. Sulphur-free, it has a more oxidized character, with honeyed, chamomile notes and less evidence of the terroir. 'Of course,' confirms Nina, 'we only sell this one to sommeliers.' It is undeniably interesting and probably makes for some creative food and wine pairings.

This is a domaine run with confidence and style, producing delicious expressions of Pouilly Fumé.

Domaine Michel Redde et Fils

Saint-Andelain

www.michel-redde.fr

The Redde family have been grape growers in the Pouilly area since 1630, but the first family member to put wine in a bottle was Michel Redde. He started the domaine in its current form in 1966, cannily choosing to site his winery and tasting room on the N7, known as the

Spot the difference: Sébastien Redde and his silhouette

'Route Bleu' because it was the main road to the south of France before the construction of the motorway. Visiting the domaine today, with its wooden tables and stools carved by Michel himself, it's easy to imagine the thriving business he must have created here.

The development of the winery continued with the next generation, his son Thierry, and today the work has been taken on by Thierry's two sons, Sébastien and Romain. They farm the domaine's 40 hectares in the Pouilly area, including a hectare of Chasselas.

The winery is constructed on three levels, so that gravity can be used to greatest effect in the winemaking. 'In the cellar, we try to do as little as possible,' says Sébastien, 'we avoid all pumping and try to work as delicately as possible. The more delicately you work, the less oxygen gets into the wine and the less sulphur [dioxide] is needed.'

The wines ferment and age for the most part in large oak, although some stainless steel is also used for fermentation of the entry-level cuvées. The brothers prefer the wood for its properties of micro-oxidation and choose different thicknesses of staves according to the terroir of the wine.

The range of wines from the domaine is of consistently good quality. The Pouilly-sur-Loire is from low-yielding vines (between 10 and 25 hectolitres per hectare), which give a wine of far greater concentration than most. It is mealy-textured and savoury, with excellent balance and a light bitterness on the finish. The Pouilly Fumé wines show purity of fruit and clear, bright flavours. La Moynerie, named after the winery, is a rich, saline wine that represents all the terroirs of the domaine. Other cuvées are terroir-specific. Les Champs des Billons vines are 30 years old and grow on Portlandian limestone. Yields are low at 45 hectolitres per hectare (the base yield for Pouilly Fumé is 65) and the grapes are hand-harvested. Les Bois Saint-Andelain comes from 30-year-old vines planted on flint, giving a highly mineral, smoky wine that is powerful and pure. 'Flint gives spice, persistence and flinty aromatics,' says Sébastien. From the estate's oldest vines, the brothers make Majorum. Created by their grandfather and only made in the best vintages, the name is individually inscribed by the brothers on each bottle in homage to him. Michel's real calling was as an artist, and his presence is still felt throughout the domaine, which is hung with his many paintings.

The brothers open old bottles of Majorum that show the ageing potential of these low-yielding, carefully made wines. And their argument that Pouilly-sur-Loire Chasselas is a wine to age is made most persuasively when they serve their Gustave Daudin wine. At 17 years old, it has gentle

Flint soils in Pouilly Fumé

aromas of smoke and rose petals and a rich palate that recalls an old Pinot Gris or Gewürztraminer, but it is still fresh and lively. 'The flint soils reinforce the acidity and the wines last a long time,' says Sébastien.

The most exciting wine of the domaine is Barre à Mine, a wine that comes from a former flint quarry. The name – 'Crowbar', in English, and featuring a label that shows a silhouetted Sébastien driving a crowbar into the ground – reflects the effort that was required to develop this site. Sébastien pulls out his phone to show dramatic footage of them dynamiting the rock before planting the vines straight into the broken rock at a density of 10,000 plants per hectare. It must have been backbreaking work. The vines are less than a decade old, but already the fruit is showing the extraordinary quality of this place. Super-mineral and spicy, the wine shows the power and heat of the rock, but it is also balanced and quite restrained, with delicious length – a fantastic and unique wine that will only get better as the vines age.

Other Pouilly Fumé producers to try

Domaine des Rabichattes (including wines in IGP Côtes de la Charité), Tinel-Blondelet, Château de Tracy, Domaine de Ladoucette, Domaine les Ormousseaux (see p. 291), Domaine Tabordet (see p. 262).

OTHER APPELLATIONS OF THE CENTRE LOIRE

Menetou-Salon

The vineyards of Menetou-Salon are dotted across the rolling countryside that lies between the outskirts of Sancerre and the edge of the city of Bourges. Compared to Sancerre the hills are lower and the vines grow mostly at altitudes of less than 300 metres. This means the grapes are faster to ripen and suffer from less stress. There is also a lower concentration of vines here and noticeably more biodiversity: cattle, sunflowers, sheep, forests and vines share the gentle, peaceful landscape.

The scattered vineyards closely reflect the outcrops of Kimmeridgian marls that lie along the gently descending limestone escarpment that runs east to west across the appellation. The climate is very similar to neighbouring Sancerre: semi-continental, with abundant rainfall of 800 to 850 millimetres per year. Spring frosts can be a risk for the vignerons here as in all the vineyards of the Centre Loire.

Menetou-Salon – key facts

- **AOC granted**: 1959
- **Hectares in production**: 625
- **Annual production volume**: 32,200 hectolitres/4.3 million bottles
- **Styles**: 75 per cent white, 23 per cent red, 2 per cent rosé
- **Grapes**: whites from Sauvignon Blanc, reds and rosés from Pinot Noir
- **Soils**: Kimmeridgian marls and limestone sediments

The wine styles closely resemble those of Sancerre: whites from Sauvignon Blanc and reds and rosés from Pinot Noir. Menetou-Salon wines have frank and open floral and citrus fruit, and are more forward and smoother in their youth than Sancerre wines. Most are destined for consumption in the first few years of their life, but those from good producers can age for a decade or longer. Vignerons may append the name of their village to the appellation, although the only village name commonly seen is Morogues. The villages, all in the Cher department, are: Morogues, Menetou-Salon, Parassy, Pigny, Aubinges, Saint-Céols, Soulangis and Vignoux-sous-les-Aix.

Notable Menetou-Salon producers

Domaine Pellé

Morogues

www.domainepelle.com

Visitors to the charming village of Morogues are greeted by the competing attractions of a twelfth-century church, an old stone well, and roses tumbling from every wall before the pillarbox-red shutters of Domaine Pellé come into view. This domaine is significant both in terms of quality and quantity, producing some of the best, if not the best, wines in the appellation from more than 50 hectares of vines.

The family started out making wine alongside a business selling grafted vine plants in the post-phylloxera years of the early twentieth century. The wine side of the business grew substantially due to Henry Pellé's tireless efforts to acquire land and to promote not only Menetou-Salon, which was granted AOC status in 1959, but also the qualities of the village of Morogues.

Anne Pellé, who runs the business with her son Paul-Henry, explains the special nature of the Morogues terroir from the vantage point at the top of the appellation, 350 metres above sea level, where the family has 12 hectares of vines. 'Menetou-Salon is flatter and the grapes ripen a week earlier. In Morogues, the terrain is steeper, higher and stonier, so the vine suffers more.' This is the Les Blanchais vineyard, the first terroir to be identified and bottled separately by her late husband Eric and his father Henry in 1981. A south- and south-east-facing site, it was planted in the late 1960s. It is a beautiful place; in the middle of the day, even nature is silent. Anne tells me they have to protect the grapes from the deer in the nearby forests, and with that, as if to prove her point, a deer runs across our path.

Today Henry's grandson, Paul-Henry, is in charge of the winemaking at the estate, which produces 70 per cent white wines and 30 per cent reds. Like many winemakers in the region today, he trained in Burgundy and has continued the practice of parcel selection started by his father and grandfather. The striking winery dates from 1970 and was built with an eye to the aesthetic. It was cut into the side of a hill to provide natural air conditioning. Whites ferment in tall, slim, elegant stainless steel tanks, while the reds, all hand-picked, go into conical wooden vats.

The domaine produces five reds and five whites, plus a rosé. They also make a red and white Sancerre. Anne aptly describes the first red cuvée, Les Bornés, as 'an easy wine, to drink with friends.' It is juicy and fleshy and entirely drinkable. The Morogues red cuvée is lovely, varietally pure, with depth and interest. The Les Cris wine comes from a monopole vineyard planted by Henry Pellé in 1967. This has more of everything: fragrance, structure and soft, fleshy fruit. It's an enticing wine. The white wines are easy to love, with clear, open fruit characters that can feel quite tropical in warmer vintages, with flavours of mango and pineapple, passion fruit and melon. An exception is the excellent Le Carroir. From a sand and flint terroir, this has a deliciously savoury, salty minerality.

Domaine Pellé is a reference for quality in Menetou-Salon and has a well-stocked tasting room that welcomes visitors year-round.

Domaine la Tour Saint-Martin

Crosses

www.domaines-minchin.com

Bertrand Minchin planted his first vines here in 1988. Today, he farms 17 hectares of grapes and also owns 200 hectares of cereals. Bertrand lives and works in a large, old farm at the western edge of the Menetou-Salon appellation, which was bought by his grandfather shortly after the Second World War.

Bertrand started out studying economics before changing course. He studied at the Lycée Viticole in Beaune and got his local inspiration from Eric Pellé. He didn't make wine for the first few years, but sold his grapes to larger estates like Domaine Pellé. His first vintage was 1994.

Bertrand started out making all of his wines in stainless steel before concluding that this can 'harden' them too much. He likes to have the option of micro-oxidation, he says, without leaving nature entirely to take its course. 'We went too far in the technical way of doing things, and then too far the other way,' he says, 'now I try to use *le bon sens paysan*,' otherwise known as common sense. He uses a combination of stainless steel and 400-litre oak barrels for the whites (his barrel cellar is the former stable in the courtyard); the reds all ferment and age in conical wooden tanks.

The focus of the domaine is white wines, although a little red is also made. The domaine cuvée Menetou-Salon Morogues is citrussy and floral (the latter a marker of Morogues terroir, in Bertrand's view), with silky texture and a touch of pepper. This is pure fruit, fermented entirely in stainless steel. The Fumet cuvée comes from a site on Kimmeridgian marl. Quite an upright, spicy and mineral wine, it is given breadth and structure from a year's ageing in barrique. A contrasting wine is Honorine, named after one of Bertrand's three daughters because, he says, it matches her sunny personality. Bertrand had originally intended to blend the wine coming from this south-facing slope with that of the north-facing Fumet site but, he says, 'one plus one did not equal two.' It is a sumptuous wine, with citrussy notes of clementine and a peachy richness that is enhanced by time in oak. It is intriguing to see how these two wines age, the Fumet acquiring notes of honey and petrol while retaining its firm acidity, and the Honorine rounding out further

Bertrand Minchin with his signature Morogues cuvée

to become more generous, softer, and almost Viognier-like. The stainless steel-fermented Morogues, not a wine particularly intended for cellaring, can nonetheless age remarkably well: at 10 years old the wine was tasting fresh and vibrant; it was heading towards golden honey and truffle essences at 15 years and still going strong at 20 years old, when it becomes more savoury, with cool cellar scents and a citrus-fresh acidity.

It is a pleasure to visit and taste wines with Bertrand, a man with great passion for his wines, not to mention a seemingly bottomless cellar of old bottles to share. He also owns a second estate, Le Claux Delorme, where he makes good varietal Sauvignon Blanc in the Touraine and Valençay appellations.

Other Menetou-Salon producers to try

Jean-Max Roger (see p. 270), Le Prieuré de Saint Céols.

Coteaux du Giennois

The Coteaux du Giennois appellation stretches for 50 kilometres along 14 communes that are almost all on the right bank of the Loire. The soils are of three main types: Kimmeridgian marls, siliceous clays or limestone, which are sometimes extensions of the soils of Sancerre or Pouilly Fumé. There are pockets of flint to be found as well.

The climate is temperate, with a continental influence. There are wide variations in temperature, but frost risk is relatively low. White wines account for two-thirds of production, but reds and rosés are also made. Whites are from Sauvignon Blanc, and these early drinking wines can offer good value. Reds and rosés must be blends of Gamay and Pinot Noir, neither variety comprising more than 80 per cent of the blend. They are open, soft and fruity, made for drinking.

Coteaux du Giennois – key facts

- **AOC granted**: 1998
- **Hectares in production**: 220
- **Annual production volume**: 11,100 hectolitres/1.5 million bottles
- **Styles**: 68 per cent white, 25 per cent red, 7 per cent rosé
- **Grapes**: whites from Sauvignon Blanc, reds blends of Pinot Noir and Gamay
- **Soils**: limestone, siliceous clay, Kimmeridgian marls

Notable Coteaux du Giennois producers

Cave Nérot

Saint-Père

www.cave-nerot.fr

Spend any time at Cave Nérot, in the little village of Saint-Père, and you will soon start to feel a part of the community. In 2016, after losing most of her harvest – and all of her export sales – to frost, Julie Nérot decided to keep things local. Now 80 per cent of her wines are sold directly from the cellar door and, judging by the steady flow of customers passing through the door on the day I visited, she knows them all by name: a couple of Dutch actors (regulars, apparently), a locally based Englishman, an elderly woman arriving with a large container to fill *en vrac* (all gone – she left with bottled wine instead) … it's all in a day's work for the seemingly indefatigable Julie. 'A lot of people come here to buy, including three generations of one family,' she tells me, in between opening bottles, pouring wine, fetching cases, washing glasses and preparing (and eating) lunch with me. It might sound frenetic, but the place has more of the feel of a rustic local wine bar than anything else, with its solid flagstone floor, simple tables and chairs and cool jazz suffusing the air. Nobody is in a hurry, and the customers chat and taste.

Julie is a fifth-generation winemaker who took over from her father in 2011 at the age of 23 ('We love each other, but we couldn't work together,' she says) and has been forging her own path ever since. She employs one worker, but otherwise does everything herself. She farms 7 hectares, all close to the winery, making white wines from Sauvignon Blanc and red and rosé blends from Pinot Noir and Gamay.

Saint-Père is at the southern extremity of the Coteaux du Giennois appellation: from here, the hills of Saint-Andelain and Sancerre are visible in the distance. The vineyards face south and the grapes ripen easily. 'We also have unproductive clones here,' says Julie, 'which accelerates ripening [because each vine has fewer grapes to ripen]'.

The wines are justifiably popular. The Sauvignon Blanc is fresh and varietally pure. The rosé, a 60:40 blend of Gamay and Pinot Noir, is pale and strawberry-scented. It is elegant, fragrant and gulpably easy to drink. Julie says, 'People think rosé is simple, but it's the hardest to make; everything needs to be done very quickly.' It must be all the harder here, because she has no temperature control in the winery. 'The settling of the juice is the most difficult,' she admits, 'but it does mean

Julie Nérot taking a break

the wine doesn't taste as "cleaned" as some rosés do.' She harvests early in the morning, 'for better quality and colour.' Her red wine, a 60:40 blend of Pinot Noir and Gamay (the opposite blend to the rosé), is made in part using whole bunches of grapes from older vines, all harvested by hand (the grapes for the white and rosé wines are harvested by machine). This is another aromatic wine, with floral notes. Elegant and fresh, with flavours reminiscent of raspberry coulis, it is an uncomplicated delight: well-judged, modern and expressive. The Gamay vines were planted by her grandfather, Julie says, more than 60 years ago, and the Pinot Noirs by her father, some 35 years ago.

The welcoming Cave Nérot tasting room is open all year and includes a shop selling everything from local ceramics to terrines and organic homewares. Julie also serves platters of cheese and charcuterie to enjoy while tasting her wines.

Domaine Les Ormousseaux

Saint-Père

Riandri Viser and Clément Jolivet are a young couple who seem destined to be together. Riandri visited Sancerre following her oenology studies and fell madly in love with the place. So much so, in fact, that on returning to her native South Africa she adopted a cat that she christened Sancerre. She returned to Sancerre five years later, and visited Clément's family's winery, Pascal Jolivet, but the pair did not meet. Then, Clément went to work in South Africa, at Klein Constantia. Eventually, they found each other, and then found their way back to Sancerre to begin their life together (with the cat, of course).

They have taken over the former Domaine Hubert Veneau, now renamed Domaine des Ormousseaux. Starting in 2021, they immediately began converting the domaine to organic viticulture and rejuvenating the vineyard: Riandri explains that nearly 50 per cent of the vines were missing.

Located in the southern part of the Coteaux du Giennois appellation, close to the border with Pouilly Fumé, the estate includes land in both appellations. When fully replanted, they will have 14 hectares of Coteaux du Giennois and 14 of Pouilly Fumé, although at the time of writing only 7 hectares are in production in each appellation. Their first vintage was 2021. The wines are gently pressed and fermented at cool temperatures in stainless steel, then aged on fine lees for a year, with a small portion of the wine going into oak for ageing. The Coteaux du Giennois white is exuberantly fruity, while the red, a minority of production, from Pinot Noir and Gamay, is light, fruity and gluggable. The Pouilly Fumé, from sand and flint soils, is light, smoky, textured and interesting, more linear in style than the Coteaux du Giennois Sauvignon.

From a young couple with energy and determination to produce wines of quality, this is a promising new venture.

Other Coteaux de Giennois producers to try

Domaine de Villargeau, Vignobles Berthier, Florian Roblin, Benoit Chauveau, Domaine Poupat & Fils.

Quincy

The story of Quincy is one of an appellation rising phoenix-like from the ashes – or rather, the frosts that all but destroyed the livelihoods of its winemakers in the spring of 1991. The appellation was on the brink of disappearing, but the Quincy vignerons were spurred into action to save their historic AOC, which was among the first in France to be classified, in the initial classification of 1936. They have worked together ever since to create a truly dynamic appellation, sharing winery and vineyard equipment. Over the years, they have invested together to buy anti-frost towers, fans that circulate the icy air to mitigate the effects of the deadly plummeting temperatures in spring. The result is that in this densely concentrated wine-growing area, 90 per cent of the vines are fully protected against frost. In 2021, for example, when many parts of the Loire (and indeed France) had their future harvest wiped out by the prolonged and severe late spring frosts, the vignerons of Quincy were able to produce a normal crop.

The Quincy winemakers say they have no trouble selling their wines. Prices have risen in recent decades and these wines are now seen as a good-value alternative to Sancerre. It's easy to see why. As a monovarietal appellation with a single wine style, Quincy wines are readily understood and of reliable quality. A bottle of Quincy will deliver crisp, dry Sauvignon Blanc. The vineyards lie on a plateau overlooking the Cher river, to the west of the city of Bourges, and the vines grow on terraces of sand and gravel over clay-limestone subsoils. Other than frosts in spring, the climate here is relatively warm and dry.

Quincy – key facts

- **AOC granted**: 1936
- **Hectares in production**: 350
- **Annual production volume**: 19,000 hectolitres/2.5 million bottles
- **Styles**: 100 per cent dry white
- **Grapes**: Sauvignon Blanc
- **Soils**: predominantly sandy with gravel and silt

Quincy producers to try

Domaine du Tremblay, Domaine de la Commanderie, Domaine Bigonneau.

Reuilly

A passing visitor to Reuilly could be forgiven for not realizing this is a wine-growing area. The flat and somewhat featureless landscape is dominated as far as the eye can see by cereal crops, punctuated by windmills. Many of the wineries continue to grow cereals such as wheat, rapeseed and barley as well as wine grapes.

But despite the evidence to the contrary, vines do grow here, close to 300 hectares in fact, producing red, white and rosé wines. The vines grow mostly on the banks of the Arnon river, to the west of Quincy and close to the small town of Reuilly. A minority of the vineyard area is on the left bank of the Cher, close to the vineyards of Quincy. The vineyards are in a narrow strip and all are close to watercourses that feed into the main river. The vines grow on shallow limestone slopes and old alluvial terraces of sandy clays and gravels.

As with neighbouring Quincy, there is a collective spirit among the growers here, many of whom share their materials and winemaking facilities in an informal, cooperative style arrangement. Like Quincy whites, those from Reuilly must be entirely made from Sauvignon Blanc. Considering the short distance between the two appellations, the styles are markedly different. The Reuilly wines have a rounder and more floral character than those of Quincy, which have a more familiar Sauvignon pungency and a pronounced acid backbone. It could be said that the Reuilly wines are a little more relaxed in style. Reuilly reds must be 100 per cent Pinot Noir. These tend to be very light in colour and body, the best with attractive fragrance. But perhaps the most interesting wines produced in Reuilly are the so-called *vins gris*, rosés made principally from Pinot Gris (although Pinot Noir may also be used for these wines). This is a local speciality, a soft and tender wine with a pale, onion skin colour and attractive, fleshy fruit that is one of the unique wines of the Loire.

Reuilly – key facts

- **AOC granted**: 1937 (whites) and 1961 (reds and rosés)
- **Hectares in production**: 292
- **Annual production volume**: 13,800 hectolitres/1.8 million bottles
- **Styles**: 56 per cent white, 24 per cent red, 20 per cent rosé
- **Grapes**: whites from Sauvignon Blanc, reds from Pinot Noir, rosés from Pinot Gris and Pinot Noir
- **Soils**: limestone marls and sandy gravel terraces

Reuilly producers to try

Denis Jamain, Domaine de la Commanderie, Domaine Bigonneau.

Châteaumeillant

This tiny appellation claims to be the most central in France (although nearby Saint-Pourçain growers might make that claim themselves. Vines have been cultivated here since the fifth century, and the vineyard reached its zenith between 1830 and 1860, when it was recorded that there were 1,200 hectares planted with Gamay and Pinot Noir. The vineyard was soon wiped out altogether when phylloxera arrived in the area, and was gradually replanted at the beginning of the twentieth century. Today more than 500 hectares of land are classified in the appellation, although only a small area is under production. The growers are keen to encourage more vignerons to the area, citing the potential for growth and the low investment needed to purchase land: twenty times less than in Sancerre and at least five times cheaper than any of the other appellations of the Centre Loire.

The location of the Châteaumeillant vineyards is somewhat unusual, having both the influence of the Atlantic Ocean to the west and the continental influence of the Massif Central to the south. This means there is more rainfall than further north, with warmer summers but colder winters. The soils are on the border of the Paris Basin and the Massif Central, and comprise predominantly sand and sandy clay over subsoils of sandstone, micaschists and gneiss.

The wines of Châteaumeillant are mainly reds, gentle and charming blends of Gamay and Pinot Noir designed to be enjoyed in the short term.

Châteaumeillant – key facts

- **AOC granted**: 2010
- **Hectares in production**: 70
- **Annual production volume**: 1,300 hectolitres/170,000 bottles
- **Styles**: 68 per cent red, 32 per cent rosé
- **Grapes**: Gamay-based, with up to 40 per cent Pinot Noir; Pinot Gris also permitted up to 40 per cent in rosés
- **Soils**: predominantly sandy siliceous soils over sandstone and micaschists

Châteaumeillant producers to try

Domaine Roux, Domaine Lecomte, Domaine Goyer.

8

AUVERGNE: MOUNTAIN WINES

The appellations of the upper Loire look and feel quite different from the rest of the Loire Valley. Far from the gentle riverside slopes and fine chateaux of the UNESCO-classified Loire, here is a more rugged and remote terrain with steep-sided cones, black basalt and granite soils, and ruins where castles used to be. Even today it is a region difficult to access without a car. The remains of the mountains of the Massif Central provide a stunning setting for the vineyards, most of which are at altitudes well above anything downstream. The vine is rather scarce here today, yet this used to be a vast wine-growing area. The region has endured a chequered history of boom and bust, both victim and beneficiary of events elsewhere.

From the early part of the seventeenth century, a local trade existed between the vineyard regions of the upper Loire, and eventually these wines found their way to Paris. This was aided by the development of canals, and improvements to the navigable upper stretches of the Allier river, principally to allow for the transport of coal from the south. Mariners were happy to supplement their cargo with wine. The vineyard had previously served the local community, but by the middle of the eighteenth century the region enjoyed a thriving trade, with regular departures from the Allier ports to take these wines to the capital.

The fortunes of the Auvergne vineyard faltered with the development of the Orléans wine industry further north. With ready access to wine at a much closer distance to Paris, merchants only ventured further south to buy wine if the harvest was poor in Orléans, or the wines were significantly cheaper elsewhere. Essentially, the wines from the Auvergne were

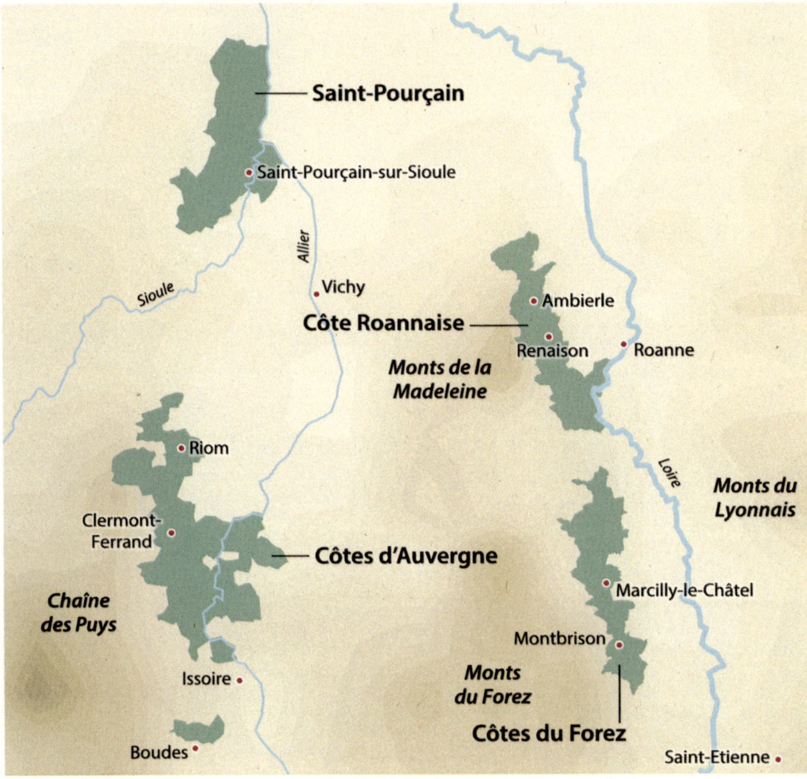

Map 11: The Auvergne

regarded as a second-best option. Even then, they were usually blended with wines from other parts of France prior to sale.

The Auvergne did experience a revival in its fortunes as a result of a widespread outbreak of powdery mildew that affected vines in the south of the country in 1855. Suddenly, wine was in short supply and Auvergne came to the rescue. Plantings in the Puy de Dôme area were recorded as 22,000 hectares in 1816 but had reached 40,000 hectares by 1862. Demand for Auvergne wines continued to boom when the next crisis – phylloxera – ravaged the vines in the south. But it was only a matter of time before the insect would travel north and wreak its devastating effects throughout the Auvergne, just like everywhere else. Although some 20,000 hectares were replanted in the period post-phylloxera, the Auvergne was once again left without a ready market for its wines and quickly declined.

Today, however, there is new impetus in the form of the Loire Volcanique growers' association. Created in November 2019, it includes

48 winegrower members from the four appellations, Saint Pourçain, Côtes d'Auvergne, Côtes du Forez and Côte Roannaise. The climate is semi-continental here, with low rainfall. Wine styles are influenced by both the Loire and its parallel tributary, the Allier, as well as the altitude of the vneyards and the effects of the surrounding mountains. Realizing they have a unique identity within the family of Loire appellations, the growers have come together to promote their story and their wines. Rather than repeating the past by falling victim to fortune or fashion, they are finding an identity of their own. No longer a source of cheap wine for blending, these small appellations are creating distinct wines of character. Whites from Tressallier and reds from Gamay Saint-Romain are particularly notable local specialities. Gamay Saint-Romain is planted almost exclusively here, and is preferred by the locals for its upright habit and small berries compared to the better known Gamay of Beaujolais. The high altitude of these vineyards means that the grapes ripen up to two weeks later than those in Beaujolais (a mere 80 kilometres to the east). The wines are lighter in alcohol and body, often with a distinctive spiciness. There are also two IGPs that cover this area, Urfé and Puy de Dôme, where some interesting varieties and blends that fall outside the appellation rules can be found.

Auvergne – key facts

- **Hectares in production**: 968 (2 per cent of the Loire's vineyards)
- **Annual production volume**: 34,600 hectolitres/4.6 million bottles (1 per cent of Loire output)
- **Style**: 48 per cent red, 31 per cent white, 21 per cent rosé
- **AOCs**: Saint Pourçain, Côtes d'Auvergne, Côte Roannaise, Côtes du Forez

SAINT-POURÇAIN

Here, they call it *le bocage Bourbonnais*, a rolling landscape of smallholdings, with little fields divided by hedgerows and an abundance of trees and wild vegetation. Even within the appellation, vineyards are far outnumbered by land devoted to the creamy-white Charolais beef cattle. Travel here in spring and you will be met with field after field of poppies, cornflowers, meadow buttercups and flowering clovers. The cows eat well here and the scent of the fields is intoxicating. In this part of France, polyculture is still common: families raise cereal crops and animals as well as vines.

Saint-Pourçain – key facts

- **AOC granted**: 2009
- **Hectares in production**: 540
- **Annual production volume**: 20,000 hectolitres/2.7 million bottles
- **Styles**: 57 per cent red, 31 per cent white, 12 per cent rosé
- **Grapes**: whites – Chardonnay, Tressallier (Sacy), Sauvignon Blanc; reds – Gamay and Pinot Noir (Auvernat); rosés – Gamay and Pinot Noir
- **Soils**: granitic soils to the west, clay-limestone slopes, sandy gravels

This is the heart of France, practically the very centre of the country. The vineyards of southern Burgundy are roughly the same distance away as those of the Centre Loire in Sancerre and the grape varieties grown in Saint-Pourçain are more closely allied to Burgundy than anywhere else. But if it feels far from the more populous and obviously 'Loire' vineyards further north, the winemakers here have no doubts where their loyalties lie. 'We are not far from Burgundy here, but we are very much from the Loire,' says Eric Nesson, of Les Terres d'Ocre.

The vineyards lie in a 30-kilometre north–south strip that covers 19 communes to the west of the river Allier, which itself is around 70 kilometres west of the Loire. The two rivers flow in a parallel course northwards until the Allier joins the Loire close to Nevers. Altitudes here are higher than further downstream so, despite the fact that the vines are several hundred miles south of Sancerre and the better-known vineyards of the Centre, the wines retain the cool-climate character of their more northerly cousins. The vineyards are at 200 to 400 metres above sea level. Red and rosé wines qualifying for the appellation Saint-Pourçain must be made from 80:20 blends of Pinot Noir and Gamay, where either grape can be dominant. Whites are from Chardonnay and Tressallier, the latter a local grape originating in Burgundy (where it is known by its official name, Sacy, also the name of a village south of Chablis). Tressallier is said to derive from a contraction of 'trans-Allier', a reference to the fact that it was transported from Burgundy, across the Allier river. Similar to Chardonnay in flavour, it ripens later and is more productive, with higher levels of acidity and lower alcohol. While these characteristics were considered inferior to those of Chardonnay in the past, a changing climate has renewed the appeal of Tressallier. And

Charolais cattle are a common sight in Saint-Pourçain

the winemakers of Saint-Pourçain are proud to have a grape variety to call their own. Whites may also contain up to 10 per cent of Sauvignon Blanc, though it does not appear to be a popular choice to use it.

Soils are sandier in the north of the appellation, next to the Allier, with clay subsoils. Further west are some parcels with granite. The heart of the appellation is a clay-limestone band that runs down the centre of the land area.

Notable Saint-Pourçain producers

Clos de Breuilly

Cesset

www.closdebreuilly.com

Acquired in 2020 by former banker and renewable energy entrepreneur William Tay-Pamart in partnership with his friend Ambroise Demonceaux, who currently still works for Apple in London, this domaine consists of an 8-hectare enclosed vineyard (*clos*) that was originally established in 1880. Wine was made at the property until 1990, when the owner decided to rent out the vines, and Tay-Pamart took the opportunity to return to his roots for his next big project. He is from Vichy and his wife from nearby Moulins. He learned about the property from fellow vigneron Denis Barbara (see p. 301), who described it to him as 'a sleeping beauty that you are going to wake up'. The domaine was an attractive proposition: 30- to 60-year-old Pinot Noir, Gamay, Chardonnay and Tressallier, all eligible for organic certification from the first vintage since the vines have been farmed organically since 2009. Keeping things in the family, Tay-Paymart teamed up with Denis Barbara's son Brian, who followed a similar path to his father by studying and working in Burgundy before returning home. The first vintage was vinified at Grosbot Barbara, but since 2021 vinification has been at Clos de Breuilly. The style of the white wines shows clear Burgundian influence, or perhaps intent, the most successful of the early wines for me being the Barnabooth cuvée, a smoky, toasty 70:30 blend of Chardonnay and Tressallier. The reds from Pinot Noir and Gamay show promise. One to watch.

Domaine des Bérioles

Cesset

www.lesberioles.com

This family business is within spitting distance of the church in the little village of Cesset. And with the domaine's summer wine bar installed directly in front of it, the Sunday congregation can move seamlessly from worship to *apéro*. How very civilized. Winemaking is a relatively recent venture in this family business, run by Jean Teissèdre, his sister Sophie and her husband Jérome Roux. Jean looks after the commercial side of the business, while Jérome and Sophie manage the winemaking. But as Jean says, 'There's a spirit of conviviality here; everyone knows how to do everything.' The small team includes a sommelier, Sylvain, who runs the tourism and tasting side of things. This is a growing enterprise that includes vineyard walks and candlelit dinners in the vines.

Jean's father bought the domaine in 1976 as purely a cereal operation. Some of the land was within the Saint-Pourçain appellation, however, and in 1989 he and his wife decided to plant a few hectares of vines. Little by little the vineyard area grew and in 2011 it was the turn of the next generation when Jean and his sister took over from their parents.

They now grow vines on 18 hectares and produce 12 different wines, mostly from individual parcels and including a couple of varietal wines from the Tressallier grape, which Jean considers their calling card. He describes the grape as 'saline, floral and mouthwatering, with flavours of citrus and fennel', and expresses his frustration that the Saint-Pourçain appellation laws restrict its use to only a portion of a blend. His varietal

Tressallier, Tressaille, is a convincing argument for vinifying it alone, a savoury, flinty mouthful with a lingering saline finish. They also make a very good sparkling *brut nature* from the variety, using the traditional method and ageing the wine for 12 to 15 months on lees. As there is no sparkling appellation here the wine sells as Vin Mousseux de Qualité Supérieur. Also outside the appellation is La Chabanne, a 100 per cent Pinot Noir from the village of Saulcet, whose soils have a very high limestone content. This is an early-drinking but highly elegant red wine that undergoes a long, soft 'infusion' to extract colour and flavour without any hard tannin or jamminess.

Chardonnay, Tressallier and Pinot Noir grapes grow on clay-limestone soils, and the Gamay on granite. The family's cereal business continues alongside wine growing, and Jean points out fields of sunflowers and quinoa across the road from his prized Tressallier vines. They also grow chickpeas, wheat and lentils. It's an impressive operation, and everything is certified organic. The wines are reliably good: well made, clean and crisp, with no excess. 'We like linear, vertical wines,' Sylvain the sommelier tells me. And as I leave, Jean hands me two bags of home-grown flour. I couldn't have been more delighted – and the flour is good, too.

Domaine Grosbot Barbara

Cesset

www.domaine-grosbot-barbara.com

A visit to Grosbot Barbara is, I learned, something of an unknown quantity. As somebody later told me, 'You know what time you arrive but you never know what time you'll leave.' As I pulled up at the house shortly before midday, Denis and Monique Barbara enquired if I'd be staying for a snack after my visit. When I finally managed to drag myself away it was close to 5 pm. Meantime, the Barbara version of a snack included the local speciality of *pompe aux grattons* (a home-baked brioche-type bread studded with pieces of fried duck skin), fresh oysters, smoked salmon, a salad of greens, ham, tomatoes, potatoes and wild garlic from the garden, numerous regional cheeses made variously from the milk of cows, sheep and goats, a home-made mint pannacotta (even the mints are local here: Vichy pastilles, a bicarbonate of soda aspirin lookalike, were invented by a local chemist for their digestive properties) and, of course, coffee. I left the pair still happily *à table* under the wisteria, listening to a nightingale singing in the trees and surrounded by the spoils of an extensive dive into their cellar.

A small sample of lunch with Monique and Denis Barbara

Denis Barbara does not cut the figure of a typical Loire vigneron, even if he describes himself as 'a pure product of Saint-Pourçain'. Clad in black, with an upturned collar, dark sunglasses, skinny jeans and flowing silver hair, he looks more eighties rockstar than winemaker, but is all boyish enthusiasm as we jump in the car to visit his latest project, which he calls 'our Côte Rôtie', or simply, 'La Côte'. The vineyard in question is quite a sight to behold: a steep, south-facing clay-limestone hillside of almost 9 hectares that he painstakingly acquired over 21 years from 26 different owners. The land, planted with vines a century ago, had long since been abandoned – clearly, there are easier places to grow grapes. The ongoing project has involved a huge exercise in tree clearance and terracing the slope. Initial plantings of Pinot Noir and Gamay are bearing fruit for vinification, with the first vintage in 2024. Inspired by the similarly steep and south-facing Treilles vineyard in Anjou (see p. 94), brought back from abandon by Jo Pithon and now part of Ivan Massonnat's Domaine Belargus, Denis Barbara and his partners at Clos de Breuilly and Domaine Nebout are planting seven varieties here, including Pinot Noir, Tressallier and Gamay. It's an ambitious vision. 'It's my revolution,' he says, 'I'm coming back to where our grandparents were.'

Meanwhile, the other wines he makes are worth seeking out. From 8.5 hectares across three sites, Barbara uses his Burgundy-acquired training and experience to produce fine, Chardonnay-based whites and smooth reds from heavily Pinot Noir-dominant blends. He describes his approach as 'grandmother's cooking', involving slow, patient and gentle infusion at low temperatures. 'It's like a gentle simmering,' he says, 'If I tried to extract colour right away, I'd lose it.' He also makes Brut Zéro, a traditional method sparkling wine, from 100 per cent Gamay. It has 18 months of lees ageing and the one I tasted had been disgorged just a few days before I visited. It was delicious, reminiscent of a Pinot-based Champagne.

Proud vineyard owner Denis Barbara

Les Terres d'Ocre

Châtel-de-Neuvre

www.lesterresdocre.com

This winery is a little out of the way in the north of the appellation, but its boldly painted red building – a converted cowshed – makes it easy to spot. If the Friday afternoon I visited was anything to go by, the genial Valérie and Eric Nesson have no shortage of local customers for their wines. The couple created the domaine in 1993 with 14 hectares, having previously sent their grapes to the local cooperative. Their smiley nephew Florent Barichard joined them in 2013. They now vinify grapes from 11 hectares for reds and 8 for whites to produce a range of wines of all colours, as well as sparkling. Some of the vines date back to the 1950s. The name of the domaine comes, unsurprisingly, from the soils, which in this part of the appellation are sandy deposits from the Allier (some 900 metres from the property) and yellow clay – the *terres d'ocre*.

The appellation wines are blends – the reds have a majority of Gamay with Pinot Noir, and the whites have a majority of Chardonnay with Tressallier. The InstanT red and white Saint-Pourçain wines are fragrant and crisp, appealingly moderate in alcohol. Tressallier accounts for the saline finish of the white, Eric tells me. They also produce a red and a

Les Terres d'Ocre

white Vin de France; the former is 100 per cent Pinot Noir and the latter 100 per cent Tressallier (thus failing to meet the appellation requirements to blend). This feels like a domaine on the ascendant. Certified organic since 2020, with everything harvested by hand, the family is pursuing some biodynamic practices as well as experimenting with ageing their wines in terracotta and sandstone jars. A concrete egg is on standby for future use.

CÔTES D'AUVERGNE

The Côtes d'Auvergne vineyard consists of disparate parcels of vines that cover an area of 60 kilometres to the north and south of the industrial city of Clermont-Ferrand. Vines grow on hillsides at between 350 and 550 metres above sea level, sometimes directly on the lava flows of the ancient volcanoes. The remarkable nature of these soils has been recognized by the creation of five specific '*crus*' within the appellation. These are otherwise (or more correctly) known as Dénominations Géographique Complémentaires (or DGCs). Wines from these five sites may have the name of the DGC appended to the basic Côtes d'Auvergne appellation. Four are for red wines only. These are: Chanturgue, Châteaugay, Boudes and Madargue. The fifth DGC, Corent, where the volcanic pumice is visible on the surface, is a permitted DGC for rosé wines only. The remainder of the appellation is characterized by clay-limestone soils (which are especially valued for Chardonnay) and basalt. Although the Côtes d'Auvergne used to number 40,000 hectares, it is now reduced to a couple of hundred, in the hands of 35 producers.

Côtes d'Auvergne – key facts

- **AOC granted**: 2011
- **Hectares in production**: 246
- **Annual production volume**: 7,200 hectolitres/960,000 bottles
- **Styles**: 65 per cent red, 20 per cent rosé, 15 per cent white
- **Grapes**: white – Chardonnay; red – Gamay and Pinot Noir
- **Soils**: volcanic and limestone-clay soils

The wild beauty of the Côtes d'Auvergne

A further 90 hectares of land is farmed for wines in the IGP Puy de Dôme. This accounts for roughly half of the production level of the Côtes d'Auvergne vineyard. Two-thirds of the wine is red and around a quarter white. Since reds must be blends to qualify for the Côtes d'Auvergne appellation and 50 per cent of the blend must be Gamay, growers wishing to produce Pinot-dominant blends (or pure Pinot Noir) must declassify their wines to IGP.

Notable Côtes d'Auvergne producers

Cave Desprat Saint-Verny

Sansac-de-Marmiesse

www.despratsaintverny.fr

The largest producer in the Côtes d'Auvergne appellation is Desprat Saint-Verny. Created in 1950, it was until 2017 the local cooperative. It was then bought by the Desprat family and is run essentially as a private cooperative. The 62 winegrowers work 180 hectares of grapes and account for around 55 per cent of the land in the appellation.

Three ranges are produced here, comprising around 20 different wines. Fusion is the entry level range, and consists of fruit-forward and approachable wines. Then comes the Eruption range, whose wines tend to be from individual parcels, aiming to express the terroir. These two ranges are usually vinified in concrete vats and see no oak. The top range is known as Magma, and signifies rare wines, often aged in oak. Unless you are at the winery or on the website, however, these classifications are of limited use since they are not typically marked on the bottles.

Côtes d'Auvergne whites are all 100 per cent Chardonnay and the wines of Saint-Verny are typically light and delicate in style. An interesting exception is 'Chardonnay 809 – the lost vineyard', named after a clone of Chardonnay which has a distinct Muscat character and is reminiscent of dry Muscat from Alsace, with its floral aromatics and dry, steely palate. The flagship cuvée, 'La Légendaire' Chardonnay, is aged for 12 months in oak barrels before bottling, and after bottling is locked in a shepherd's hut at 1,200 metres altitude for six months. The story goes that Léa Desprat's grandfather, Pierre Desprat (whose face adorns the label), would take a bottle of wine to go hunting and, after leaving a bottle one year in a shepherd's hut, found it tasted better the following year. The

Caves Desprat Saint-Verny

wine merits its flagship status and is now the subject of an annual festival when the wine is transported back down the mountain via a human chain. Another favourite is 'La Vallée des Saints Boudes', a light, inexpensive and invigorating blend of 50:50 Gamay and Pinot Noir, which has pure, crunchy cherry fruit and a mineral, peppery, slightly saline finish.

Les Chemins de l'Arkose

Montpeyroux

The logo on the label of Les Chemins de l'Arkose looks like some kind of aboriginal symbol. In fact, it is an artist's impression of Yvan Bernard, co-owner of this Montpeyroux-based winery. 'My wife is a hairdresser, but I'm not a good advertisement,' says the hirsute Yvan, who is nevertheless happy enough to feature this depiction of his face not only on his wine bottles, but also on his calf, where it is tattooed alongside a similar drawing of fellow owner, Audrey Baldassin.

Yvan set up his business 20 years ago, when it was known as Domaine Yvan Bernard. Working with Audrey Baldassin since 2014, he decided to join forces with her to create the newly branded company in 2020. They farm 13 hectares; one-third around the tiny hilltop village of Montpeyroux and nearby Sauvetat, one-third in the Corent *cru*, and one-third further north, above Clermont-Ferrand. Around 50 per cent of the production is Gamay, with Pinot Noir, Chardonnay, Syrah, Aligoté and Seibel making up the rest of the plantings.

The soils are varied here: Montpeyroux is a granite hill, Sauvetat is limestone-clay and Corent is volcanic, consisting of volcanic ash, lava and basalt, with a clay subsoil. Côtes d'Auvergne is manifestly volcanic, and nowhere is this more evident than in the *cru* of Corent. Unusually, the appellation allows only rosé here, from Gamay. Yvan's Gamay grows at 450 metres above sea level on the eastern side of this former volcano, where the ground is strewn with chunks of lava and basalt. His rosé is bright in colour and vinous, with lovely purity of flavour. More of a serious wine for food than one for sipping by the pool, this could easily keep for a year or two.

The approach to winemaking here is natural – at least as far as is practical. Wines are bottled without sulphur only if Yvan considers them to be stable. 'We only have one harvest a year,' he says, 'so we can't afford to lose it. We taste, we test, we work with a laboratory, we analyse the wine.' The wines of Les Chemins de l'Arkose are full of energy and clear fruit with lightness and balance. 'Les Dômes sur Basalt' is a joyful

Yvan Bernard in the volcanic Corent cru

expression of juicy Gamay fruit grown on basalt, with vibrant acidity. 'The most crucial thing is the date of the harvest,' Yvan informs me, 'you mustn't be afraid of a few pink grapes.' It is the kind of wine I'd be happy to drink every day. The Vieux Gamays cuvée comes from vines planted between 1940 and 1960, on granite. The granite gives firmer wines with more concentration, he tells me. The wine has gorgeous black fruit, is neither fined nor filtered and has no added sulphur. This feels like intelligent natural winemaking. Yvan works with an oenologist from Mâcon who visits the region periodically to advise a few of the more naturally-minded winemakers. It's something of a trek to get to this remote wine-growing region of the Massif Central. As Yvan says, 'We are all alone here – we have the impression of being pioneers.'

Also of note

Jean Maupertuis.

CÔTE ROANNAISE

The vines of the Côte Roannaise occupy hillsides a few kilometres to the west of the Loire at altitudes of between 400 and 500 metres above sea level. The vineyards run for 25 kilometres north to south, on volcanic and predominantly granite soils, on east-facing slopes. To the west rise the foothills of the Monts de la Madeleine. The speciality here is Gamay Saint-Romain, which is the only grape permitted in the appellation, although there are a good number of wines of every colour and multiple varieties made within the IGP Urfé.

Côte Roannaise – key facts

- **AOC granted**: 1994
- **Hectares in production**: 154
- **Annual production volume**: 5,300 hectolitres/707,000 bottles
- **Styles**: 90 per cent red, 10 per cent rosé
- **Grapes**: Gamay
- **Soils**: granite, schist, gneiss

Notable Côte Roannaise producers

Domaine des Pothiers

Villemontais

www.domaine-des-pothiers.com

Romain Paire's family has been making wine in the same place – exactly the same place – since 1610. 'We're not great adventurers,' he says. The family house dates back to the year 1000, he tells me. Romain farms 23 hectares, mostly around this house, although he has started to invest in a few local vineyards slightly further afield recently. Most of the vines grow at between 400 and 500 metres above sea level. Now, however, he tells me, 'Everything I'm planting is above 500 metres.' This is a luxury not available in most places, but it offers insurance of sorts through cooler growing conditions in an increasingly uncertain climate.

Romain took over the family estate in 2005. In his grandfather's time the family lived mostly from raising Limousin cattle and, although the focus today has switched firmly to winemaking, they still keep a few cows. The entire estate is farmed biodynamically and all the grapes are harvested by hand. The care taken in the vineyard shows in the wines, which are consistently good across a wide range of grapes and styles.

There are 12 cuvées showing individual plots, and the reds from Gamay Saint-Romain show the depth and interest this grape can produce when grown on these granitic soils. Fermentations are in concrete, wood, or amphorae, depending on the site, the age of the vines and the creative inclination of the winemaker. The entry-level cuvée, La Colline en Flamme, is from young vines. The grapes are partly destemmed and undergo a short maceration. It is a perfumed, soft and supple wine. The Domaine cuvée spends time in wood and concrete before bottling. It is a deliciously fleshy wine with meaty depth. Vinified in the same way is the juicy cuvée No.6. Instead of granite, though, these vines grow on clay. La Chapelle comes from vines planted 520 metres above sea level on green granite. It spends a year in conical wooden tanks and has rich, sweet fruit. This is a very cool terroir, says Romain, and the vines are in their prime, at 25 years of age. It's a pure pleasure to drink. Clos du Puy comes from a walled vineyard at 515 metres above sea level, a beautiful, breezy spot where the soils are of rhyolite, pink granite and quartz. An early-ripening site, it yields a wine with rounded fruit and plummy depth that needs time to open. L'Intégrale vineyard is at 460 metres altitude. This wine ages for a year in 500-litre barriques and is more marked by wood than most. It shows savoury fruit intensity, with grilled meat, liquorice and firm tannins. The pinnacle of the range is Diogène, a wine that spends eight months macerating in terracotta amphorae. The grapes come from the same site as L'Intégrale but the wine could not be more different. Aromatic and pure-fruited, with floral notes and a crisp acid backbone, it also has a wonderful saline quality and featherlight texture. The absence of oak brings the spiciness of Gamay Saint-Romain to the fore.

Romain makes several whites, from Chardonnay (an oaked and an unoaked version), a lovely Pinot Gris (Hors Piste) and – unusually – a Riesling. The Riesling – En Paradis – is a new project, the young vines producing wine that is already fragrant and varietally pure. 'We have a lot in common with Alsace here,' he comments, 'east-facing slopes with altitude – and granite, the best soils for Riesling.'

Domaine Florent Thinon

Villemontais

As I was driving through the vineyards of the Côte Roannaise with Romain Paire of Domaine des Pothiers, he pointed out a young man in the vines, identifying him as Florent Thinon. 'He is always there,' he said. When I later meet this rising star of the Loire, Florent Thinon tells me he has been in his uncle's vines since he was '*tout petit*'. Realizing from an early age that wine was his passion, he left home to study in Beaune for four years before gaining experience at Château Thivin in Beaujolais and Domaine des Bérioles in Saint-Pourçain. On returning to his native Côte Roannaise, he was taken under the wing of his neighbour, Stéphane Sérol, and worked alongside him at Domaine Sérol. By 2020, Florent was in a position to take over 3.5 hectares of his uncle's vines to set up on his own. 'I started from zero,' he says, but he already had a clear idea of the style of wines he wanted to make: 'agreeable, with some structure, and as much finesse and elegance as possible.'

Florent now farms 7 hectares of Gamay Saint-Romain and half a hectare of Chenin. He planted the latter himself in 2019, but the rest of the vines average 30 years of age. Everything is hand-harvested and sorted on *tables de trie*. The majority of the grapes are destemmed, but he leaves 30 to 40 per cent of bunches whole. By harvesting early in the morning, he ensures the grapes go into the concrete fermentation tanks relatively cool. This allows them to infuse gently before fermentation starts as temperatures rise. The grapes are macerated for 15–20 days. 'I manipulate very little,' says Florent, 'The objective is to have as little impact as possible on the tank. I do a small, five-minute *remontage* twice a day, morning and evening.' He is aiming for infusion rather than extraction to achieve the elegance and finesse he seeks. The finished wines are aged for a short period of 6 to 11 months, mostly in concrete tank, before bottling.

The main cuvée, Le Rouge, has delicious clarity of fruit and pure flavour. Crisp and elegant, it combines the generous, sunny fruit of Gamay with a refreshingly low level of alcohol (typically only 12–12.5% abv). A second Gamay cuvée, Mayençat, has darker fruit and a more savoury character. A little more austere in its youth than Le Rouge, it is made from grapes grown at a higher altitude of 550 metres on south-facing slopes of clay and granite, and is aged for 11 months, partly in 500-litre, old oak barrels, although Florent is quick to say that he is not aiming to 'add wood.'

From his first vintage in 2020, Florent Thinon has made rapid progress, doubling his land under production, creating a new white wine from Chenin and a zippy direct-press Gamay rosé. At a mere 24 years of age, he is making wines with the confidence of someone twice his age. The next step is to team up with someone even younger – his little brother, Clément.

Domaine Sérol

Renaison

www.domaine-serol.com

This high-quality family domaine in the heart of the appellation has been here since the seventeenth century. It is run by husband and wife Stéphane and Carine Sérol, who farm their 38 hectares with extraordinary dedication.

Stéphane took over the domaine in 1996 when it was only a third of the size it is today, and set about increasing the size of the vineyard by clearing unplanted hillside sites. The vines grow at altitudes of between 400 and 550 metres. The domaine is certified biodynamic and all the work is done by hand – as much through necessity as anything, because the vines are planted at a high density: up to 10,000 plants per hectare

Carine Sérol and the domaine's screwcapped rosé – an unusual sight in the Loire

(compared to the more usual 5,000). 'We can't use tractors here,' says Stéphane. When I visit in spring, he and his team are furiously busy in the vines, debudding to manage the yields. Everything is debudded, deleafed and harvested by hand. They also plant cover crops which can be ploughed into the soil to manage the moisture levels. 'There are no rules,' says Stéphane, 'just observation and adaptation.' The vines are almost all Gamay Saint-Romain and include a parcel, Les Blondins, in shared ownership with Michel Troisgros (the three-star Michelin restaurant Troisgros is less than 10 kilometres away). In addition, they farm a couple of small parcels of Viognier and Chenin.

With such a huge amount of work in the vines, it's quite a surprise to find the couple's house looking picture-perfect, a rose-clad granite building with neatly pollarded trees for shade. This used to be Stéphane's grandparents' house, explains Carine, conveniently sited next to the winery. A welcoming tasting room is open year-round.

The 'house wine', as Carine calls it, is the Eclat de Granite. This, the largest cuvée of the estate, shows savoury, bright fruit, a light grip and good length. It is fermented slowly with indigenous yeasts and ages for up to ten months in concrete tanks. This comes from several plots; most of the other wines are from single-vineyard sites. Another Gamay Saint-Romain, Oudan, is a fragrant and floral wine that comes from a south-facing slope of pink granite planted by Stéphane. It is vinified in large oak tanks with 40 per cent whole bunches, then aged in the same large wooden tank for 11 months. The vines for Les Millerands are between 50 and 95 years old. They grow on a high, east-facing granite hillside, densely packed at 10,000 plants per hectare. There is no wood used here, but the wine has the distinct spiciness of Gamay Saint-Romain, along with cherries and chocolate. It is pure, fresh and light. Perdrizière grapes grow on a soil of old, decomposed granite (the same soils are found in Moulin-à-Vent and Côte de Py in Beaujolais, Carine tells me). The soil is rich in iron oxide and manganese. This is a cool, east-facing site that gives a smooth and silky wine, fermented and aged in large oak vats. Around 50 per cent of the harvest is vinified as whole bunches. It is a delightful wine, with firm tannins, liquorice, spice, clove notes and floral aromas. From the centre of this plot comes another wine, Chez Coste, made from destemmed grapes that ferment very slowly and then age for a further six months, all in terracotta amphorae. Firm but broad, this is another expression of the Gamay Saint-Romain, and one that needs time to reveal itself.

There are also dry whites from Chenin and Viognier, an orange, amphora-macerated, textured Chenin with mineral, savoury flavours, a light, summery Gamay rosé sealed with a screwcap and a refreshing, strawberry-scented *méthode ancestrale* sparkling to enjoy from this modern and well-run estate.

Also of note

Domaine Reniteo.

CÔTES DU FOREZ

The Côtes du Forez may be small, but it has a couple of claims to fame. It both has the highest altitude vineyards of the Loire, and is the appellation closest to its source. As the growers are fond of saying, 'The Loire starts here.' The history of wine growing stretches back as far as the monastic orders of the tenth century, but its recognition as an appellation contrôlée is recent. It comprises less than 100 hectares of planted vineyards (although there could legally be many hundreds more), spread along a 45-kilometre stretch and managed by ten growers and a cooperative. The vines grow on basalt and granite at between 400 and 600 metres above sea level. A Foehn effect from the Monts du Forez to the west means that rainfall here is low.

The wines are all Gamay, in a similar style to those of the Côte Roannaise although, as president of the appellation Gilles Bonnefoy explains, the difference between the two appellations comes down to the higher altitude of the Côtes du Forez vineyards and the higher percentage of clay in the soils. Clay soils tend to give wines with more pepper and spice compared to the rounder and less acidic wines of the Côte Roannaise.

Côtes du Forez – key facts

- **AOC granted**: 2000
- **Hectares in production**: 74
- **Annual production volume**: 2,400 hectolitres/320,000 bottles
- **Styles**: 85 per cent red, 15 per cent rosé
- **Grapes**: Gamay
- **Soils**: granite, basalt

Notable producers

Les Vins de la Madone

Champdieu

www.lesvinsdelamadone.fr

Gilles Bonnefoy is a native of Champdieu, but after an early career in the commercial side of wine, he decided to get his hands dirty and start his own winery from scratch. In 1997 he started looking to rent a few vines. He says it was easy at this time to build a collection of a few old vineyards because older people who had previously sent their wine to the cooperative were keen to sell. Soon he had accumulated more land and planted vines on the steep hillsides surrounding the domaine. The name of the domaine comes from the volcano of La Madone opposite the winery, at the top of which a large, white statue of the Madonna glows in the sun and can be seen from miles around.

Gilles Bonnefoy in the high-altitude vineyards of the Côtes du Forez

The domaine now stands at 12.5 hectares and is certified biodynamic. Gilles has installed a modern winemaking facility. The vines grow on slopes that face east, south and south-east. The biggest problem they face, says Gilles, is drought. To mitigate this he would prefer to be able to plant on lower slopes with deeper soils, but this is not permitted in the appellation regulations.

The main cuvée of the domaine is Gamay sur Volcan, which grows on soils that are half volcanic and half granite with clay. All of the domaine's grapes are hand-picked and yields are low, and for this wine the grapes are destemmed. Fermented in stainless steel with native yeasts and bottled young, this is a fragrant and silky mouthful, with supple, melted tannins. It's perfect for early drinking. Mémoire de la Madone is another notable Gamay, coming from low-yielding 50-year-old vines that grow on basalt. The grapes are partly destemmed and the wine ages in sandstone amphorae. There is real depth of flavour and concentration here, with a grainy texture and a lovely stony character.

Gilles also makes a lovely Roussanne that grows at an altitude of 550 metres on the steep slope directly beneath the statue of the Madonna. It has floral fruit, a lemony tang and mouthwatering acidity. Other wines reflect individual terroirs and, sometimes, different varieties of Gamay

Roussanne vines on the volcano of La Madone

he has planted, including a *pétillant naturel* from the pink-fleshed Gamay Teinturier, called Bulles de la Madone.

Also of note

Domaine Verdier Logel.

GLOSSARY

AOC/AOP: interchangeable acronyms for Appellation d'Origine Contrôlée/Protégée. These letters or words are used on wine labels following the name of the appellation, or sometimes surrounding it, for example 'Appellation Savennières Contrôlée'.

Barrique: small wooden barrel used for fermentation or maturation, typically 225 litres or 228 litres in capacity.

Bâtonnage: the process of stirring lees into suspension during ageing of (usually white) wine. Lees-stirring can add richness to wines and creamy, mealy, savoury, umami characters.

Biodynamic viticulture: Based on the writings of Rudolf Steiner, this is an holistic method of viticulture that considers the vineyard as a self-sustaining organism. To be certified, growers must first have organic certification. Biodynamic methods include adding regular homeopathic doses of plant-, animal- and mineral-derived preparations whose applications are ideally timed according to the movements of celestial bodies, using the Biodynamic calendar. The emphasis is on promoting natural balance in the vines to enhance their resistance to pests, diseases and weather events.

Botrytis cinerea: also known as noble rot (*pourriture noble* in French), or simply botrytis. *Botrytis cinerea* is a fungus whose spores penetrate ripe grapes in the presence of humid conditions in autumn. If the wet conditions persist, the bunches will rot (this is known as botrytis bunch rot, or grey rot) and the grapes will be lost, but if the humid weather is followed by warm and dry weather – as happens in the afternoons in the great sweet wine vineyards – the fungus dries to a dark, ashy appearance. The process continues as long as these favourable conditions are in place, drying the ripe grapes and concentrating the sugars and acids within them, while imparting the unique flavours of the fungus.

Brettanomyces: often abbreviated to Brett, this is a yeast that can infect fermenting and finished wines, especially those aged in wooden barrels. Red wines are most often, but not exclusively, affected, and wines with low or absent sulphur dioxide, or with high pH levels are more at risk of infection. Brett is responsible for horsey, barnyard or sticking plaster aromas and, while it is usually considered an undesirable fault, some appreciate its presence in small doses or, occasionally, large ones.

Cap: the floating mass consisting mostly of grape skins and pips that is pushed to the surface in red winemaking by the carbon dioxide released as a by-product of fermentation.

Chaptalization: the practice of adding sugar to grapes or must prior to fermentation to increase the potential alcohol of a wine.

Clone/clonal selection: clonal material is used to allow an entire vineyard to be planted with identical characteristics. Plant breeders will select clones based on their suitability for the growing environment and rootstock, and to provide more regular and predictable results in the vineyard. The alternative is massal selection (see below).

Clos: a legally protected term meaning a walled vineyard, or a vineyard that used to be surrounded by walls. A wine name that includes the word 'Clos' must be from a site that meets these criteria, but this is not the case for the winery if it is an historic name, such as Clos Rougeard, or Clos Naudin.

Cru: commonly used to denote superior quality, very few wines in the Loire are entitled to use the term on a label. The exceptions are the Coteaux du Layon Premier Cru Chaume and the Grand Cru Quarts de Chaume, although it is a term routinely used to describe DGC wines such as those awarded in Muscadet, e.g. Clisson.

Curetage: see Esca.

Demeter: the principle certifying body for biodynamic agriculture. Another, French-based and wine-specific, certifying body is Biodyvin.

Demi-muid: oak barrel, usually with a 600-litre capacity.

Demi-sec: literally meaning half-dry but more usually referring to off-dry styles. For Chenin Blanc this typically means between 8 and 18 grams of residual sugar, but for Vouvray, for example, this is not regulated and the term is used as an indication of sweetness. Winemakers may decide for themselves if they think a wine is in the *demi-sec* style or not.

***Dénomination Géographique Complémentaire* (DGC)**: A permitted addition to a wine label that gives further information about the origins of a wine. It may also relate to certain winemaking and vineyard practices, e.g. for the Muscadet *crus* (which are not currently *crus* in the legal sense of the word).

Dosage: sweetened wine added to a sparkling wine before bottling. A wine that is labelled *non-dosé* may still have a small amount of sweetness from residual unfermented sugar.

Doux: used to describe a wine that is obviously sweet, and sweeter than *moelleux* styles.

Downy mildew: major vineyard fungal disease that spreads rapidly during damp, warm weather and can cause devastating crop loss if left untreated.

Élevage: the maturation of a wine after fermentation, often in wooden barrels.

Esca: a serious trunk disease that is spread by fungal spores entering pruning cuts in the wood of mature vines. Sauvignon Blanc vines are particularly susceptible to esca. The disease blocks the sap flow of the vine and is often fatal, although its progress can be limited or halted by removal of the affected wood, a process called *curetage.*

Extraction: the process of getting colour and flavour from the solid parts of grapes into a wine. Most commonly used during red winemaking (but also during white wines with skin contact), the choice of extraction methods will play a large role in determining the final style of the wine. Extraction takes place at various times during the winemaking process, and any time a grape's juice is in contact with its skins there is the potential to extract colour and flavour compounds, whether intentional or not. This could be at any moment from harvest throughout fermentation, maceration post-fermentation and pressing.

Fermentation: the process by which carbohydrates (in winemaking, the sugar in grapes) are converted to alcohol. Sometimes referred to as alcoholic fermentation, or primary fermentation, the latter especially used in reference to sparkling winemaking.

Foudre: general term for a large oak barrel or vat (considerably larger than a *barrique*, for example).

Franc de pied: A term sometimes seen on labels to describe a wine made from vines planted directly into the ground on their own roots instead of grafted onto phylloxera-resistant American rootstocks.

Fût: a generic term for a wooden barrel.

Green harvest: the process of picking and discarding immature grapes during the summer months to manage yields and improve the balance and ripening of the remaining bunches.

***Haute Valeur Environnementale* (HVE):** an environmental certification programme set up by the French Ministry of Agriculture in 2012. It can be applied to any form of agriculture, but has been adopted with most enthusiasm by the viticulture sector, which makes up around 60 per cent of all certifications. Wide-ranging in its scope, it considers biodiversity, cover crops and hedges, and reduced dependence on inputs, including fertilizers, agrochemicals, energy and water usage. It is less stringent than organic certification, which prohibits use of synthetic inputs.

Hectare: a unit of land measurement equating to roughly 2.5 acres.

Hectolitre: 100 litres.

Hectolitres per hectare: a standard measurement of yield, often abbreviated to hl/ha. 50 hectolitres per hectare is equivalent to approximately 3 tons per acre.

***Indication Géographique Protégée* (IGP):** a wine classification that is indicative of origin but with less stringent rules than AOC/AOP. Permitted yields are higher and there is greater flexibility on the grape varieties that can be used compared to AOC wines.

Indigenous yeasts: Also known as native yeasts (the alternative being commercial yeasts), these are strains of yeasts that occur naturally in vineyards and cellars.

Institut National des Appellations d'Origine (INAO): the French government body responsible for creating and monitoring all the rules and regulations relating to the appellation classifications. The rules for appellations are continually being revised and updated.

Layering: also known as *marcottage*, this is a vine propagation technique that involves burying the shoot of a mature vine in the ground to create a new plant. Especially popular in Vouvray, the technique enjoys a high success rate, but the new vine is at risk from phylloxera since it is not grafted onto a resistant rootstock.

Lees: the dead yeast cells and other matter remaining after alcoholic fermentation. Many white and sparkling wines benefit from spending time ageing on lees (referred to as ageing *sur lie* in French, or *sur lattes* for bottle-fermented sparkling wine). Lees have antioxidative properties, meaning wines can spend many years 'resting' on their lees, as is the case for Muscadet *crus* and the best sparkling wines. The lees impart yeasty, bready, mealy and creamy notes to the wines, and these characteristics are accentuated by lees stirring, a process known as *bâtonnage*.

Lieu-dit: An officially named vineyard (recognized in local maps, rather than just from the mind of a plot owner) that may or may not be classified in some way.

Ligérien: a term used to describe something that is typcially from the Loire. Ligérien wines are light, fresh and fruity, with crisp acidity.

Liquoreux: a sweet wine term, usually the sweetest of a style.

Malolactic fermentation: also known as malolactic conversion, malo, or simply mlf, this is a winemaking process where lactic acid bacteria transform tart malic acid into softer lactic acid. It can be provoked by creating ideal conditions for it to occur but more usually it occurs naturally after, or sometimes during, fermentation in the presence of warm temperatures and the absence of sulphur dioxide. It confers richness, buttery flavours and a fuller mouthfeel.

Marcottage: see layering.

Massal selection: a traditional method of selecting the best plants for propagation to ensure the favourable characteristics of these vines are passed on to future generations. Massal selection is often chosen to create greater diversity of plant material in a vineyard. The alternative is clonal selection (see above).

Mesoclimate: the climate within a specifed vineyard.

Méthode ancestrale: also known as *pétillant naturel*, this is a method for producing sparkling wines by harnessing the carbon dioxide that is produced during the primary (and only) alcoholic fermentation. Typically, a fermentation is started in tank and then the still-fermenting wine is transferred to bottle to complete its fermentation, leaving a wine that is dry (assuming all of the sugar has fermented), and with lower pressure than a sparkling wine that has undergone a primary fermentation to dryness followed by a second fermentation in the presence of extra added yeast and sugar.

Microclimate: the climate in a small area, such as a single vine or cluster of vines.

Moelleux: a French word that translates as 'soft', it is also a term applied to wines sweeter than *demi-sec* but less sweet than *doux* or *liquoreux*. Vouvray wines in this category typically have up to 45 grams per litre of sweetness.

Mouse taint: a poorly understood but noxious wine fault that, unlike cork taint (TCA), remains undetectable by many people (up to a third of the population). It is thought to be on the rise due to an increasing number of wines bottled with no or low sulphur dioxide addition but, as it is difficult to measure, its prevalence remains unclear. It is odourless and, initially, tasteless, detectable only when the pH of the wine changes when it reacts with the (higher) pH of human saliva. The wine then reveals a most unpleasant flavour, likened to a hamster cage or dead mouse.

Native yeasts: see indigenous yeasts.

Natural wines: a term without an agreed definition, although it is often taken to mean wines made and bottled with minimal intervention on the part of the winemaker and without recourse to additional inputs (such as yeasts, enzymes or, especially, sulphur dioxide). See box, p. 18.

Négociant: a company that mainly buys grapes or wine from others to sell on, although may also have its own vineyard holdings.

Noble rot: see Botrytis.

Oidium: also known as powdery mildew, oidium is a common fungal disease that can cause significant crop loss.

Organic viticulture: a natural method of grape growing that prohibits use of synthetic agrochemicals in the vineyard. Growers may use copper and sulphur to combat threats to the vine.

Parcellaire: a French term that refers to the practice of separating, vinfying and bottling wines according to soil type to create site-specific wines.

Pétillant naturel: see *Méthode ancestrale.*

pH: a measure of the concentration of acidity in a liquid. The lower the pH, the more acidic the wine. Some grape varieties have naturally lower pH than others. Climate affects the pH of grape juice as well, and warmer vintages will usually result in higher pH values. Wine pH is usually between 2.9 and 4.

Phylloxera: A sap-sucking insect that was accidentally introduced to Europe from the United States towards the end of the nineteenth century, destroying the entire French vineyard. It is still considered a major risk in vineyards the world over, but vines are now protected by grafting the *vinifera* plant onto rootstocks of resistant American vines.

Pigeage: Also called punching down in English, this is the submerging of the cap of grape skins during red wine fermentation to ensure even colour and flavour extraction. There are various tools available to do this, from mechanical to the human foot. An alternative to *pigeage* is pumping over or *délestage,* where wine is pumped or drained off from the bottom of the tank and returned via the top.

Pourriture noble: see Botrytis.

Powdery mildew: see Oidium.

Racking: the process of removing a wine from its sediment (such as grape skins and yeast residues) from fermentation onwards. It may be done during fermentation to aerate the must before returning to the tank, or at the end of fermentation before *élevage* or bottling.

Residual sugar: sweetness (measured in grams per litre) remaining in a wine once it is bottled. This may be natural, unfermented grape sugars or, in some cases, a combination of grape sugar and sugar added in the form of chaptalization at the start of the fermentation. For sparkling wines sweetened just before bottling, the sweetness is known as *dosage.*

Rootstocks: The solution for rebuilding the French vineyard after phylloxera was to graft the vulnerable *Vitis vinifera* grape varieties onto American rootstocks resistant to attack. Since the louse travels through the soil and attacks the roots of the vine, the *vinifera* variety is thus protected from damage. However, the process of grafting can weaken the vine and the graft can be a point of entry. Some believe there is a loss of the 'true' nature of the original grape variety, so a number of growers risk planting some of their vines directly into the soil (see *franc de pied*).

Sec: a term used to describe dry wine in terms of not only absence of residual sugar but also perceived sweetness. For most grape varieties *sec* will mean less than 3 grams of sweetness. But for Chenin Blanc wines this figure will often be higher because the high acidity of this variety will make wines taste drier. For example, in Vouvray and Touraine, *sec* is used to define wines with under 8 or 9 grams of residual sugar.

Sulphur: an antioxidant and antimicrobial substance that is widely used in viticulture (in the form of copper sulphate) and winemaking (in the form of sulphur dioxide).

Terroir: a borrowed French word lacking a succinct English translation that is used to describe the overall growing environment of a vine. More loosely defined it also includes other elements unique to a wine, such as the conditions for *élevage* or the hand of the vigneron.

Traditional method: sparkling wine production method, also known as the Champagne method (or *méthode champenoise)*, involving two fermentations. The first fermentation takes place in the usual way in a large tank or other vessel. The wine is then bottled with additional yeast and sugar and sealed, provoking a second fermentation in the bottle. Here the carbon dioxide produced by the fermentation is trapped inside the bottle, creating the mousse. The wine then rests on its lees for a period of time (determined by appellation law and the desire of the winemaker) before disgorgement, or *dégorgement*, where the yeast lees are removed from the wine before it is resealed and sold, usually with a sweetening *dosage* added at the bottling stage.

VDQS: Short for *Vin Délimité de Qualité Supérieur*, this was a category of wine created to bridge the gap between table wine and AOC wines. The category was abandoned in 2011, when many of the wines classified as VDQS were awarded AOC status instead. The VDQS category has been replaced by IGP.

Vigneron: A term without a direct translation, so the original French is frequently used. It describes someone who grows grapes and makes wine. In English we might say winegrower. Usually only the masculine form of the word is seen in English, but the French would always use *vigneronne* for a female winegrower.

Vin de copains: literally, a wine for friends, wines described in this way are in a similar category to *vin de soif*: early drinking, not too serious, wines for pleasure (as indeed they are sometimes called: *vins de plaisir*).

Vin de garde: a winc suitable for ageing.

Vin de soif: 'Soif' means thirst in French, which makes this term pretty self-explanatory. It is used to describe wines for drinking, rather than thinking about. A summer version, used especially for rosé wine, is *vin de piscine*: swimming pool wine.

Vin gris: a term that sounds so much more pleasant in French than its English translation of 'grey wine', *vin gris* is a rosé that is light in colour, typically with a pale, onion skin tint. Originally called *gris* to distinguish it from more deeply coloured rosés, now that pale rosés are everywhere, it could equally just be called rosé.

Whole bunch fermentation: where bunches of grapes are placed directly into a fermentation vessel without destemming.

APPENDIX I: VINTAGES 2004–2024

A WORD ABOUT VINTAGES

In a northerly winegrowing region such as the Loire Valley, the vintage really counts. Even if growing seasons have been warmer in recent years, and average temperatures are certainly rising, there is still much variability and unpredictability in the weather from year to year. The Loire being such a spread out region, there can also be significant differences from place to place and between one wine style and another.

When thinking about vintages, it is worth bearing in mind that one of the most appealing features of Loire wines is that most of them are approachable from an early age and do not require years in bottle before opening. The Loire offers freshness and drinkability across all its wine styles, now more than ever, as vignerons are creating wines that are a more faithful reflection of the terroirs where their grapes grow. Instead of trying to make wines that need time, they are increasingly producing wines for drinking now, emphasizing the Loire's naturally bright, often delicate, cool-climate fruit.

Even if most wines can be drunk young, however, the best of all styles do benefit from ageing. The top reds from Chinon, Bourgueil and the appellations around Saumur are made to be aged, even if they can often be approached within five years. Cabernet Franc matures more quickly than Cabernet Sauvignon and, after just a few years, can already be showing attractively softened fruit and tannins. Despite this, the wines can continue to develop remarkably well, especially in a good vintage; the great 1989s are still showing well today, for example.

Most of the Pinot Noirs of the Centre region can be enjoyed young, but the best of Sancerre (and a few from Menetou-Salon) can easily age for a decade. Less obvious, perhaps, is the fine ageability of the top Sauvignons from the Centre region. Growers from Sancerre and Pouilly Fumé are insistent on the need to wait for several years for their wines to lose unwanted varietal characters and to reveal instead the uniqueness of their terroirs. I have tasted 30-year-old Sauvignons from the Monts Damnés vineyard in Sancerre that were still delicious and perfectly fresh. Sauvignons from Touraine are, with the occasional exception, early-drinking styles. Most of the Loire's Gamay is also intended for early consumption, but the top cuvées from the rewardingly spicy Gamay Saint-Romain wines in the Auvergne will improve with an additional few years in bottle.

Perhaps a surprise contender for ageing is Muscadet. Usually thought of as a wine for immediate consumption, Muscadet – in particular from the *crus* – ages effortlessly.

But the most exceptionally ageworthy variety of the Loire is Chenin Blanc, although again, in most of its incarnations, bottle ageing is not required. The best wines, however, across all its still wine styles, will certainly improve and evolve with age. The top dry wines of Savennières, Anjou Blanc and Saumur can be austere in their youth and often benefit from a few years to fill out in bottle, especially if they have spent extended time ageing in oak. The best will improve for decades. The Loire's sweet Chenins are easy to enjoy when young but can age almost indefinitely, adding layers of complexity to their youthful frames.

2024

A vintage whose grapes are not picked at the time of writing, but which promises to be a challenging one for growers, given the extremely wet growing season from spring through to midsummer. It will be a small crop and easier for whites than reds.

2023

A perfect early season with excellent fruit set and potential quantity turned into a rain-soaked summer and autumn, with inevitable and often significant crop losses due to mildew and rot in the western part of the region. Growers who were able to select fruit effectively made attractive, early drinking wines. Elsewhere rain led to dilution and lower acidities than desirable, notably in the Sauvignons of Sancerre and the Central vineyards, where yields were also extremely high.

2022

This was a uniformly warm and very dry vintage. After the frosts of 2021 the vines compensated, giving generous crop loads. There were some high heat spikes that led to vines shutting down for a period in some places. The wines have excellent ripeness levels and many are very good. Some whites show atypically tropical flavours and generally have gentler levels of acidity, but on the whole are well balanced. Where the naturally high yields were not controlled, there is dilution in places. Care was needed in extracting tannins in the red wines, as the extended period of drought thickened the skins of the grapes. In some instances, tannins are showing in the Chenins as well. Overall, though, it is a very good vintage for all styles, with many delicious wines to drink or keep.

2021

Known as the year of the most devastating frosts most had ever seen (especially in Muscadet, but throughout the Loire and indeed all over France), this was a small vintage. Beyond the spring frosts, the weather continued to be dismal in 2021. In the words of one grower, 'there was no summer'. The wines are described by many as *ligérien* – in other words, of a classically Loire style – with fresh acidity and modest fruit and alcohol. Despite the trials of the year, 2021 has produced elegant, restrained whites and many are very good wines, especially the Melon and Sauvignon wines, which are more consistent than the Chenins. Reds from 2021 are less successful and can be quite hard work, with a challenging combination of high acidities and herbal fruit. The best growers, however, produced delicate, crisp wines for early consumption. There are a few rare and excellent sweet botrytized Chenins from this vintage too.

2020

This was the third warm summer in a row, with good yield and ripeness levels, and many attractive wines were made. Some whites, especially Chenins, show higher than usual alcohol levels due to rapid end-of-season ripening. Acidities are relatively low and the overall quality is good. There are some successful reds, too, also with high alcohol levels at times, although the excessive summer heat shut down the vines and delayed ripening in other cases, causing unripe tannins. Overall, this is a good vintage, though.

2019

Despite a diminished crop due to spring frosts, 2019 turned out to be a very good to excellent vintage in most places, the reduced yields allowing for good concentration. Wines are well balanced and expressive, with high ripeness but also very fresh acidity, which is a delicious combination for both reds and whites. Chenins are round and ripe with good freshness. The best Sauvignons are balanced, although in some cases alcohol levels are a little high. Reds are perhaps the most successful: deeply coloured, fresh and aromatic, with lovely natural balance.

2018

This was a happy vintage for vignerons across the region, resulting in good quality and high yields. The first of three warm vintages, the 2018s show high levels of ripeness and softened acidity levels. Whites are atypically ripe and rich due to the warm summer and absence of rainfall. Sauvignons are exotic in profile, and Chenins reliable, if less racy than usual. Some are a little unbalanced by high alcohol levels, and flavours are definitely heading towards the tropical, but many are plump and delicious. Red winemakers were justifiably happy and there are very good, ageworthy reds from 2018, with ripeness levels well above the average for the region. It was a rare, relaxing vintage for the vignerons of the Loire.

2017

This was a year of terrible spring frosts, notably in Savennières, but also in Muscadet and Vouvray. Nevertheless, the remaining small crop of wines, both red and white, are of good to very good quality.

2016

The frosts were devastating in 2016, especially in Muscadet, and crop levels were low throughout the Loire. Summer was wet, but towards the end brightened up to ripen grapes across all styles, producing a small harvest of good quality.

2015

This was a good vintage with a warm and dry summer, although it turned out to be cooler than the 2018–2020 trio of vintages that were to follow. Absence of frost gave growers a generous crop and the summer heat gave early drinking, fruit-forward whites and ripe, full-bodied reds.

2014

This was a fairly cool year with a long growing season that produced excellent, mineral and ageworthy wines of all styles in the classic *ligérien* style.

2013

The 2013 growing season was a cool one that was interrupted by hail in June in Vouvray that put paid to most of the crop (although nearby Montlouis escaped the worst of the frost). The vintage was overall better for whites, which are fresh and mineral. The reds show decent ripeness and modest levels of alcohol.

2012

A cool vintage that was very good for early ripening whites such as Muscadet and the Sauvignons of the Centre, which show lively acidity, fragrance and complexity. The weather was less favourable for reds, some of which show unripe flavours. Heavy rains in September made for a challenging vintage for Chenin.

2011

This was a wet year characterized by disease and rot. The Chenins in Anjou fared better than grape varieties elsewhere. There were some fine botrytized wines made in the Layon.

2010

This was a good year for many of the whites of the Loire. The long, cool season gave Muscadets of concentration and minerality. Vouvray and Montlouis were affected by harvest rains, but the later season weather was perfect for botrytis in the Layon Valley. September rains slowed ripening of red wine grapes, but growers who held out for ripeness were rewarded with classic Cabernet Francs.

2009

This was a generous, dry and ripe vintage producing reliably good red and white wines in a richer style. Muscadets are ageworthy, with good levels of acidity. Chenins are generally good and there are some worthy sweet wines, although botrytis was scarce. The unusual ripeness led to warm alcohols in some styles, such as Savennières and Sauvignons from the Centre. Reds show good concentration of fully ripe fruit.

2008

A universally cool year, with much of the summer overcast, leading to bright, classically styled wines. The sun appeared in September and October to give sufficient warmth to ripen the crop. Cool Sauvignons and Chenins can be found from 2008, but there is an absence of sweet wine due to rain later in October.

2007

This was a cool and often wet vintage that produced fresh, herbal red wines, good Muscadets and some notable botrytized Chenins.

2006

Not a vintage of note. September rains meant selective picking was needed. Some good Chenins in dry and *demi-sec* styles.

2005

A good vintage across the board, with a fine summer and early harvest that was rather hot for the whites, giving quite rich and luscious Sauvignons and a good quantity of excellent botrytized Chenins. This was a particularly successful vintage for reds, which show ripeness and balance.

2004

A moderately good vintage, with irregular performance in different regions. There were successful wines from Muscadet, Savennières and the Layon Valley. Reds are firm, a result of alternately warm and cool conditions across the growing season.

Notable older vintages

2002 (whites), 1996, 1990, 1989, 1988

APPENDIX II: LOIRE PRODUCTION, FIVE-YEAR AVERAGES

Five-year averages: hectares in production and volume produced by region 2019–23

Region	Hectares in production	Percentage of Loire total	Volume produced (hl)	Percentage of Loire total
Nantais	7,603	13.5	302,655	10.9
Anjou, Saumur, Haut-Poitou	18,828	33.4	989,235	35.7
Touraine	12,899	22.9	574,222	20.7
Centre	5,971	10.5	330,708	12.0
Auvergne	968	1.7	33,076	1.2
Total AOC	**46,269**	**82.0**	**2,229,896**	**80.5**
IGP	4,826	8.6	257,155	9.3
Other	5,308	9.4	282,843	10.2
TOTAL LOIRE	**56,403**	**100.0**	**2,769,894**	**100.0**

Source: Service Economie et Etudes, InterLoire 2024

APPENDIX III: 2023 WINE PRODUCTION IN THE LOIRE VALLEY

Appellation	Hectares in production	Output (hl)	Average yield (hl/ha)	Colour/style
The Nantais				
Coteaux d'Ancenis	145	7,546	52	white, red, rosé
Fiefs Vendéens	355	18,513	52	rosé, red, white
Gros Plant du Pays Nantais	329	20,312	62	white
Muscadet	1,190	64,701	54	white
Muscadet Coteaux de la Loire	92	4,239	46	white
Muscadet Côtes de Grandlieu	205	9,270	45	white
Muscadet Sèvre et Maine	4,720	209,941	44	white
Muscadet Sèvre et Maine Crus Communaux	143	5,239	37	white
Total Nantais	**7,179**	**339,761**	**47**	

Appellation	Hectares in production	Output (hl)	Average yield (hl/ha)	Colour/style
Anjou, Saumur and Haut-Poitou				
Anjou	1,414	72,920	52	red, white, sparkling
Anjou-Villages	138	6,660	48	red
Anjou Brissac	57	2,464	43	red
Anjou-Coteaux de la Loire	13	403	31	white
Bonnezeaux	23	472	21	white
Cabernet d'Anjou	5,871	352,015	60	rosé
Coteaux de l'Aubance	115	3,132	27	white
Coteaux de Saumur	5	110	21	white
Coteaux du Layon	1,037	30,872	30	white
Coteaux du Layon Chaume	17	362	21	white
Coteaux du Layon Villages	63	1,012	16	white
Coulée de Serrant	7	227	33	white
Crémant de Loire	3,659	271,376	74	white, rosé
Haut-Poitou	101	6,371	63	white, red, rosé
Quarts de Chaume	9	54	6	white
Rosé de Loire	701	40,456	58	rosé
Rosé d'Anjou	1,520	102,343	67	rosé
Saumur	2,543	170,055	67	sparkling, red, white, rosé
Saumur-Champigny	1,413	82,798	59	red
Savennières	145	5,620	39	white
Savennières-Roche aux Moines	18	495	27	white
Total Anjou, Saumur, Haut-Poitou	**18,869**	**1,150,217**	**61**	

Appellation	Hectares in production	Output (hl)	Average yield (hl/ha)	Colour/style
Touraine				
Bourgueil	1,182	52,630	45	red, rosé
Cheverny	702	39,008	56	white, red, rosé
Chinon	2,376	96,216	40	red, rosé, white
Coteaux du Loir	84	3,553	42	red, white, rosé
Coteaux du Vendômois	91	4,571	50	red, white, rosé
Cour-Cheverny	72	3,714	52	white
Jasnières	76	3,638	48	white
Montlouis-sur-Loire	421	21,016	50	sparkling, white
Orléans	39	1,068	28	red, white, rosé
Orléans-Clery	10	318	31	red
Saint-Nicolas-de-Bourgueil	1,078	60,818	56	red, rosé
Touraine	3,709	230,320	62	white, red, sparkling, rosé
Touraine-Amboise	81	3,968	49	white, red, rosé
Touraine-Azay-le-Rideau	42	1,495	36	white, rosé
Touraine-Chenonceaux	164	9,654	59	white, red
Touraine-Mesland	70	2,904	42	red, rosé, white
Touraine Noble-Joué	35	1,886	55	rosé
Touraine-Oisly	38	2,307	60	white
Valençay	218	12,415	57	white, red, rosé
Vouvray	2,193	136,486	62	sparkling, white
Total Touraine	**12,681**	**687,985**	**54**	

Appellation	Hectares in production	Output (hl)	Average yield (hl/ha)	Colour/style
Centre Loire				
Châteaumeillant	75	1,666	22	red, rosé
Coteaux du Giennois	233	13,645	59	white, red, rosé
Menetou-Salon	647	43,916	68	white, red, rosé
Pouilly Fumé	1,412	91,091	65	white
Pouilly-sur-Loire	24	1,493	61	white
Quincy	353	26,240	74	white
Reuilly	304	17,988	59	white, red, rosé
Sancerre	3,053	206,741	68	white, red, rosé
Total Centre	**6,101**	**402,780**	**66**	
Auvergne				
Côtes d'Auvergne	239	7,661	32	red, rosé, white
Côte Roannaise	170	6,318	37	red, rosé
Côtes du Forez	80	2,609	33	red, rosé
Saint-Pourçain	523	25,003	48	red, white, rosé
Total Auvergne	**1,012**	**41,591**	**41**	
Total AOC	**45,842**	**2,622,334**	**57**	**white, rosé, red, sparkling**
Total IGP	**4,746**	**337,640**	**71**	**white, red, rosé, sparkling**
Vin de France	**5,690**	**417,300**	**73**	**white, rosé, red**
TOTAL LOIRE	**56,278**	**3,377,274**	**60**	**white, rosé, red**

Source: Statistiques des Récoltes des Vins de Loire, Service Economie et Etudes, InterLoire 2024

BIBLIOGRAPHY

(undated) *Le Marché des Vignes.* Available at: https://www.safer.fr/app/uploads/2024/05/2024-PDT2023-04-Vignes.pdf (Accessed: 05 August 2024).

Asselin, C. and Girault, P. (2017) *Le val de loire, terres de chenin*. Doué-en-Anjou: Éditions les Caves se rebiffent.

Avia, K., Schneider, C., Onimus, C., et al. (2023) 'The French grapevine breeding program ResDur: state of the art and perspectives', *IVES Conference Series, GiESCO 2023*. Available at: https://ives-openscience.eu/33918/ (Accessed: 11 August 2024).

Boulay, T. (2020) *Les Terroirs Sancerrois: Un héritage géologique, Culturel et Immatériel.* Eyrolles.

Brouard, E. and Jenni, A. (2021) *La Loire et ses vins: Deux mille ans d'histoire(s) et de commerce*. Paris: Flammarion.

Dorey, A., 'Ces Cuvées Sur Lesquelles tout amateur de Grands Vins Blancs Devrait Miser, à moins de 30 euros', *Le Figaro Vin,* 15 March 2024. Available at: https://www.lefigaro.fr/gastronomie/le-muscadet-est-le-nouveau-montrachet-ces-cuvees-sur-lesquelles-tout-amateur-de-grands-vins-blancs-devrait-miser-a-moins-de-30-euros-20240315 (Accessed: 22 September 2024).

Feiring, A. and Lepeltier, P. (2017) *The Dirty Guide to Wine: Following flavors from ground to Glass*. New York, NY: The Countryman Press.

Goujard, A. and Saverot, D. (2024) 'Louis-Benjamin Dagueneau: "Depuis 2018, nous touchons les limites du cépage sauvignon", *La Revue du Vin de France*, 682 (July/August), pp. 8–11.

Ladet, N. (2021) *Innovative grape varieties to revitalise vineyards.* INRAE Institutionnel. Available at: https://www.inrae.fr/en/news/

innovative-grape-varieties-revitalise-vineyards (Accessed: 11 August 2024).

Lascelles, A., 'The New Juju of Anjou Wine', *Financial Times*, 28 August 2024. Available at: https://www.ft.com/content/8eea9a40-34d4-4b89-9d4a-7c0f1cc60f68 (Accessed: 22 September 2024).

Lepeltier, P., Rigourd, P. and Bardin, J.-Y. (2022) *Vignerons de Loire: 170 portraits de vignerons en liberté*. Angers: Fervel édition.

Maisonneuve, P. (1925) *L'Anjou, Ses Vignes et Ses Vins*. Angers, France: Imprimerie du commerce.

Maisonneuve, P. (1926) *L'Anjou, Ses Vignes et Ses Vins. Deuxième Partie, le Vigneron Angevin*. Angers.

Maltman, A. (2020) *Vineyards, rocks, and soils: The wine lover's guide to geology*. New York: Oxford University Press.

Neethling, E., Barbeau, G., Quénol, H. and Bonnefoy, C. (2012) 'Change in climate and berry composition for grapevine varieties cultivated in the Loire Valley', *Climate Research*, 53(2), pp. 89–101.

Peydro, L. (ed.) (2017) *Chinon: Voyage au pays de Rabelais*. Paris: Gallimard.

Raduget, N. and Yung, E. (2022) *Histoire des vins de l'aoc touraine*. La Crèche: Geste éditions.

Rigaux, J. et al. (2008) *Pouilly-Fumé: Jewel of the Loire Valley*. Clémencey, France: Terre en Vues.

Robinson, J., Harding, J. and Vouillamoz, J. (2012) *Wine grapes*. Penguin Random House.

Voss, R. (1995) *Wines of the Loire*. London: Faber and Faber.

Williams, D., 'Why the Loire is France's most exciting wine region', *Guardian*, 23 April 2024. Available at: https://www.theguardian.com/food/2024/apr/23/why-the-loire-is-frances-most-exciting-wine-region (Accessed: 22 September 2024).

Wilson, J.E. (1998) *Terroir: The role of geology, climate, and culture in the making of French wines*. London: Mitchell Beazley.

Useful websites

www.agriculture.gouv.fr/les-chiffres-cles-de-la-haute-valeur-environnementale-hve – French government website with legislative information about, for example, HVE (High Environmental Value) certification

www.federationviticole.com – Fédération Viticole Anjou Saumur

www.inao.gouv.fr – for information relating to INAO regulations
www.jasnieres.fr – Jasnières wines
www.loire-volcanique.com – the Loire Volcanique growers' association
www.muscadet.fr – Nantais wines
www.musee-boissons.com – for general history and information about French appellations
www.vigneronsdelatouraine.com – Touraine wines
www.vins-centre-loire.com – representing the wines of the Centre Loire appellations
www.vinsduhautpoitou.com – Haut Poitou wines
www.vinmethodenature.org – natural wine certification body
www.vinsvaldeloire.fr – InterLoire website

ACKNOWLEDGEMENTS

This book is the product of hundreds of encounters and thousands of tastings, and it could not have been written without the help and support of many people. I offer my heartfelt thanks to all the wine growers who have helped with my research in any way. Thank you for your generosity in sharing your wines, your time and your stories with me. For those who feature within the pages of this book, I hope I have done justice to your words and your wines. I needed no convincing that Loire wines are wonderful with food; nevertheless I greatly appreciate all those who demonstrated this beyond doubt by sharing wonderful meals paired with their wines. How fortunate I am.

There are too many individuals to mention here, but thanks especially to Evelyne de Pontbriand and Ivan Massonnat for their hospitality in Savennières, and to Gilles Bonnefoy for helping me out in the wilds of the Côtes du Forez. Thank you to Patrick Baudouin for his insights into Anjou, Chenin Blanc, and much more. I am grateful to François and Vincent Lieubeau for their warm welcome and efficient help in Muscadet and I am indebted to Vincent for taking the time to share his fascinating research into the terroirs of Muscadet. I am thankful to Lionel Gosseaume for the wealth of information he provided on Touraine, and to Frederik Wilbrenninck for inviting me to taste the historic vintages of Moulin Touchais (twice). Thanks also to Amélie Néau for her help in organizing visits in Saumur-Champigny.

There is a lot of ground to cover in visiting the vineyards of the Loire, and the task of managing this was made considerably easier with the help of individuals representing grower organizations. Thank you to Muriel Roudaut and the Syndicat de Chinon, François Bouteille in

Vouvray, Camille Roblin and Anne Burchett in Centre Loire, François Robin in Muscadet and Marion San Bartolomé in Montlouis-sur-Loire. Representing the Loire Volcanique growers, Aurélie Soubiran was most helpful – thank you, Aurélie. I am grateful to Alexis Trentesaux, Hélène Kermorvant and everyone behind the scenes at InterLoire for providing multiple opportunities for me to taste many of the region's wines. Particular thanks are due to Fanny Gautier for the invaluable statistical information used throughout this book. I should also like to thank Etienne Neethling at INRAE for his insights into climate change and clonal research, and Sam Schrock at IFV for showing me the world of ResDur varieties.

I am grateful for the significant input I have received along the way from professional colleagues and friends. Massive thanks and love to Rebecca Gibb MW, who has shared most of this amazing journey with me and helped me in more ways than I can count. This is a better book for our many hours of joint visits, tastings, discussions and debate. I would not have done it without you. I should also like to express my gratitude to Richard Kelley MW, Sam Harrop MW, Martin Hudson MW, Jim Budd, Chris Kissack, Chris Hardy, and Jon David Headrick. From the Classic Wine Library, a big thank you to Sarah Jane Evans MW for her enthusiastic endorsement of my initial idea, and to Rebecca Clare and Hermione Ireland for their patience and skill in getting the book to its final beautiful state with no drama. I am most grateful to Kerr MacRae for his professional advice. To my family and friends, thank you for bearing with me. And finally, to my husband Geoff, eternal thanks for your unwavering support, patience and love.

INDEX

Note: producers are indexed by surname. Where family members are discussed only under the producer name, they are not indexed separately.